TRIALOGUES AT THE
EDGE OF THE WEST

Terence McKenna (l), Ralph Abraham (c), and Rupert Sheldrake (r).

RALPH ABRAHAM
TERENCE McKENNA
RUPERT SHELDRAKE

TRIALOGUES AT THE EDGE OF THE WEST

CHAOS, CREATIVITY, AND THE RESACRALIZATION OF THE WORLD

FOREWORD BY
JEAN HOUSTON

BEAR & COMPANY
PUBLISHING
SANTA FE, NEW MEXICO

LIBRARY OF CONGRESS CATALOGING-IN-PUBLICATION DATA

Abraham, Ralph.
    Trialogues at the edge of the West : chaos, creativity, and the resacralization of
the world / by Ralph Abraham, Terence McKenna, and Rupert Sheldrake.
        p.   cm.
    ISBN 0-939680-97-1
1. Cosmology. 2. Panpsychism. 3. Consciousness. 4. Gaia hypothesis.
5. Chaotic behavior in systems. 6. Religion and science—1946-
I. McKenna, Terence K., 1946-   . II. Sheldrake, Rupert. III. Title.
BD511.A27 1992
113—dc20   91-44753
                                                                                                CIP

Copyright © 1992 by Ralph Abraham, Terence McKenna & Rupert Sheldrake

Bear & Company, Inc.
Santa Fe, NM 87504-2860

Cover design: Robert Aulicino
Cover & text fractal images: Art Matrix © 1990
Interior design: Chris Kain
Text renderings: Marilyn Hager Biethan
Author photo: Tom Galloway © 1991
Editing: Ralph Melcher & Gail Vivino
Typography: Buffalo Publications
Printed in the United States of America by R.R. Donnelley

1 3 5 7 9 8 6 4 2

To Esalen

CONTENTS

Chapter 1: Creativity and the Imagination 3

The new evolutionary cosmology. The regularities of nature as
evolving habits. The basis of cosmic creativity. The cosmic imagination as a higher-dimensional attractor drawing the evolutionary
process toward itself. The Omega Point. Imagination welling up
from the womb of chaos. Psychedelic experience and the mind
of Gaia. Gaian dreams and human history. Dark matter as the
cosmic unconscious.

Chapter 2: Creativity and Chaos 23

The chaos revolution. Chaotic attractors as eternal mathematical
realities. Indeterminism in nature. Chaos and the evolution of order.
Form in the cooling process. The organizing fields of nature as
related to mathematics and the cosmic imagination. Mathematical
models. Attractors, attraction, and motivation. The freezing of
information in crystals and in written language. The primacy of
spoken language and abstraction.

Chapter 3: Chaos and the Imagination 43

Chaos in Greek mythology. The myth of the conquest of chaos.
Fear of chaos and the suppression of the feminine. The partnership
society and the rise of patriarchy. Seasonal festivals of the repression
of chaos, and the creation of the unconscious. The inhibition of
creativity and its relation to global problems. The Eleusinian mysteries. Creativity and Christology. Plans for the recovery of chaos
and the imagination. The significance of the chaos revolution.

ILLUSTRATIONS

ACKNOWLEDGMENTS

We are grateful to Esalen Institute, especially Nancy Kaye Lunney and Steve Donovan, for the hospitality and encouragement that made this project possible and to Paul Herbert and Marty Schrank for their excellent recordings. We are also deeply indebted to Jill Purce for leading the chants with which we began each trialogue. Finally, we would like to acknowledge the extraordinary support of the Bear & Company staff, in particular, Barbara Hand Clow for her excellent editorial suggestions; Ralph Melcher for his sensitive editing and supportive enthusiasm; Gail Vivino for her fine tuning and careful copyediting; Barbara Doern Drew for her creative approach to production; and Chris Kain for her aesthetic and appropriate book design.

FOREWORD

When I was quite a young child, my father, a comedy writer, invited me to go with him to deliver a script to the ventriloquist Edgar Bergen, whose weekly radio show he was writing at that time. Bergen's chief dummy, Charlie McCarthy, was one of the best-loved characters in radio comedy and was featured in many movies as well. He was also my very dear friend, and when he was sitting on Bergen's knee, we would have many sprightly and madcap conversations.

When Dad and I entered the open door of Bergen's hotel room, we found him sitting on a bed with his back to us, talking very intently to Charlie and then listening with evident wonder and astonishment to Charlie's answers. Unlike in the radio programs, there was no flippancy here, no "in-on-the-joke" sarcasm. Indeed, one got the impression that Bergen was the student, while Charlie was quite clearly the teacher.

"What are they doing?" I silently mouthed at my father. "Just rehearsing," he mouthed back. But as we listened to what Bergen and Charlie were saying, we soon realized that this was no rehearsal for any radio program we ever knew about, for Bergen was asking his dummy ultimate questions like "What is the meaning of life? What is the nature of love? Is there any truth to be found?" And Charlie was answering with the wisdom of millennia. It was as if all the great thinkers of all times and places were compressed inside his little wooden head and were pouring out their distilled knowings through his little clacking jaws.

Bergen would get so excited by these remarkable answers that he would ask still more ultimate questions: "But, Charlie, can the mind be separate from the brain? Who created the universe, and how? Can we really ever know anything?" Charlie would continue to answer in his luminous way, pouring out pungent, beautifully crafted statements of deep wisdom. This rascally faced little dummy dressed in a tuxedo was expounding the kind of knowing that could have come only from a lifetime of intensive study, observation, and interaction with equally high beings. After several minutes of listening spellbound to this wooden Socrates, my father remembered his theological position as an agnostic Baptist and coughed. Bergen looked up, turned beet red, and stammered a greeting. "Hello, Jack. Hi, Jean. I see you caught us."

"Yeah, Ed," my father said. "What in the world were you rehearsing? I sure didn't write that stuff."

"No rehearsal, Jack. I was talking to Charlie. He's the wisest person I know."

"But, Ed," my father expostulated, "that's your voice and your mind coming out of that cockeyed block of wood."

"Yes, Jack, I suppose it is," Bergen answered quietly. But then he added with great poignancy, "And yet when he answers me, I have no idea where it's coming from or what he's going to say next. It is so much more than I know."

Those words of Bergen changed my life. For I suddenly knew that we contain "so much more" than we think we do. In fact, it would seem that in our ordinary waking reality we live on the shelf in the attic of our selves, leaving the other floors relatively uninhabited and the basement locked (except when it occasionally explodes). I also knew that I had no other choice but to pursue a path and a career that would discover ways to tap into the "so much more" of deep knowledge that we all carry in the many levels of reality and nested gnosis within ourselves.

In reading *Trialogues at the Edge of the West* I found myself once again eavesdropping on an extended conversation of ultimate questions and far-reaching answers. This time, however, the principals are no dummies, unless one thinks of the very cosmos as their ventriloquist. Ralph Abraham, Terence McKenna, and Rupert Sheldrake are among the brightest and most thoughtful men alive on the planet today. Still, the mystery of intelligence ignited and the calling forth of incendiary visions remains. These thinkers quicken in each other a remembrance of things future as well as things past. They evoke from one another a new treasure trove of ideas that could keep us all thinking for the next hundred years—so much so that I find myself wondering, To whom and to what am I listening? What is this book? A concerto of cosmologists? An atelier of thought-dancers? A marching society of the ancient Gnostic order of metaphysical inebriates? Or have the spirits of Thomas Aquinas, Peter Abelard, and Saint Bonaventure come back to run rampant through the polymorphous parade of intellectual possibilities felt and known by each of these fellows?

Whatever else this is, *Trialogues* is surely a mine field of mind probes, a singular sapient circle of gentlemen geniuses at their edgiest of edges. Their metaphors alone would leave a Muse in a muddle.

Meeting yearly and, more recently, publicly with each other at Esalen, they raise through their conversations the rheostat of consciousness of themselves and their listeners. They cut loose from whatever remains of orthodox considerations and become minds at the end of their tethers, who then re-tether each other to go farther out in their speculations. In so doing, they have figured out how to achieve one of the best of all possible worlds: the sharing of mental space and cosmic terrains over many years of deep friendship and profound dialogue.

The questions they pose each other are of the sort that the Hound of Heaven brings barking to our heels. Here is a sampling: Are the eternal laws of nature still evolving? Is there a realm beyond space and time that grants the patterns and the conditions for creativity, organization, and emergent evolutionary process—or does the universe make itself up as it goes along? Are the causes of things in the past or are they in the future? Is there some hyperdimensional, transcendental Object luring us forward? Is history but the shadow cast backward by eschatology? Are we humans the imaginers or the imagined; or is history in some way a co-creation—an unsettled, chronically evolving, funky partnership between ourselves and the hyperdimensional Pattern Maker? Are the visionary vegetables our potentiators and our guides; and is theobotany the key to it all? Is chaos merely chaotic, or does it harbor the dynamics of all creativity? What is the connection between physical light and the light of consciousness? How do we breach our fundamental boundaries so as to enter a new phase of the human adventure?

Let the reader be warned that this is a curiously initiatory kind of book, one that serves to recreate the landscapes and inscapes of our culture, our science, and ourselves. The participants in these trialogues have in their own ways striven to green the current wasteland by the remarkable range of their human experience as well as by their depth of thought. This they have done by personally engaging in more levels of reality, investigating the range and depths of the ecology of inner and outer space, and bringing back rich travelers' tales of their discoveries.

This series of trialogues is a living testament to the fact that we are living in times during which our very nature is in transition. The scope of change is calling forth patterns and potentials in the human brain/mind system that as far as we know were never needed before. Things that were relegated to the unconscious are moving into consciousness. Things that belonged in the realm of extraordinary experiences as well as ideas of the nature and practice of reality are becoming ordinary. And many of the maps of the psyche and its unfolding are undergoing awesome change. This is not to say that there are not perennial things about the deep psyche that will always remain generally true. But our ways and means of reaching them are through routings never known before. The principals in *Trialogues* are leading the ways in exploring the new routings. They show us how discontinuities and multiple associations of the old tribal societies have again become important. Chaos theory becomes critical in understanding the way things work. We must look for flow patterns rather than linear cause-and-effect explanations. Resonance—both morphic and interpersonal—has become far more important than relevance. The world is now a field of colossal busybodyness with quanta of energies effecting everything simultaneously. And with this resonance, nothing is truly hidden anymore. This is why this is also a book of secrets revealed—on every page; no, more, in every paragraph. *Caveat emptor.*

The rapidity with which ideas are here offered, plumbed, and then potentiated—the speed and passion with which myths and symbols are presented and then rewoven into new tapestries of the spirit—serves as witness to the acceleration of the psyche in our times. The human psyche is one of the great forces of reality as a whole. It is a thing that bridges what may appear to be separate realities—that is, it is a great force of nature, it is a great force of spirit, and it is greatest of all as the tension that unites the two. Now this psyche may be moving toward phase-lock breakout—that is, the jump time of the psyche—manifesting as many different singularities of itself as it moves toward convergence and transition. This means it is moving at stupendous speed past the limits most people have lived with for thousands of years into an utterly different state of mind. The contents of the psyche are manifesting at faster and faster rates—a dreamlike reality in which it is difficult to tell anymore what is news and what is drama—or, for that matter, what is myth and what is matter.

We live in a surround of electronic stimulation that extends to all tribes, nations, peoples, realities, and the Earth herself through every one of us. We leave out nobody. Everybody has to participate eventually, however impoverished and unseen they may be at the moment. We have been returned electronically and in fractal waves of multicultural convergence to a tribal world of instantaneous information and dialogue.

Trialogues at the Edge of the West is this worldwide phenomenon writ small. Within it, as in the world we now live in, realities come as thick and fast as frequencies. We are constantly sitting at the shore absorbing the frequency waves of these realities, peoples, experiences, and energies all the time. This absorption, I maintain, is changing all the patterns of the ways in which we are composed and, by extension, the ways in which we now have to orchestrate and conduct this new composition. We live in chaos, which we may have created in order to hasten our own meeting with ourselves— that is, to blow down the old structures that no longer sustain us. In our lifetimes, the great sustaining cultures have moved from agri-culture to factory culture to technoculture to omniculture. And people like Abraham, Sheldrake, and McKenna are emergent apologists for this omniculture. This they can be because of their appreciation of myth.

All over the world myths have risen to conscious popularity because we can no longer understand the dreamscapes of our everyday waking life. The myth is something that never was but is always happening. It serves as a kind of DNA of the human psyche, carrying within it the coded genetics for any number of possible evolutionary and cultural paths we might yet follow. This is quite possibly why these fellows couch their language in mythic cadence. They know they are on to Some Thing and perhaps, even, to Some One, so their mouths are metaphored and become full of the blood. The authors are on the verge of telling the new and larger Story. You see and read at another level. *Trialogues* becomes a text of "Ceremonies at the Edge of History," an Eleusinian mystery play in which we are invited to join the three celebrants to sing as Pindar did when he reflected on the ancient mysteries: "Blessed are they who have seen these things. They know the end of life, and they know the God-given beginnings."

Jean Houston
Pomona, New York
March 1992

Jean Houston, Ph.D., is the author of a dozen books including The Hero and the Goddess, The Search for the Beloved, *and* The Possible Human. *She is a philosopher, psychologist, cultural historian, and well-known seminar leader and international consultant on human development. She is also the director of the Foundation for Mind Research.*

PREFACE

After living in Nainital in the Indian foothills of the Himalaya, at thirty-six years old I returned to California. One day, as I was standing on a street corner in front of the Santa Cruz post office in white robes waiting for a ride, a car stopped and Doug Hanson, a friend whom I had not seen in a year, said, "Get in, I have someone for you to meet." Having no other agenda, I got in. He left me in front of a frame house on Carlton Street in downtown Berkeley. I went up one flight of stairs to a little attic, where Terence McKenna was stooped over a terrarium, studying a dead butterfly and living mushrooms. We started talking, roving over a vast landscape everywhere familiar to us both, and hours passed.

Over the course of the seventies we evolved a pattern of relating, including a minimum of chitchat, dinners, and hours of dialogue followed by sleep. Our talks on philosophy, mathematics, and science created a space between us for mutual exploration and discovery, which diffused into my professional work. A paper called "Vibrations and the Realization of Form," published in 1975 by Erich Jantsch, came out of this space.

In September 1982, a routine visit to Terence was interrupted by a phone call announcing the arrival of Rupert Sheldrake at the bus station. I had recently read his book A New Science of Life, which had caused a stir in England, and found it extraordinarily compatible with my own thought. When we picked up Rupert, he entered effortlessly into the mental space Terence and I had created over a decade, as his presence stretched the space into an equilateral triangle.

Through the eighties we explored and extended our trialogue, making many thrilling discoveries as a bonded triad, self-conscious of our trinity, synergy, and partnership. Occasionally, at various conference and performance platforms, our activity emerged into public view. Eventually, with the encouragement of Nancy Kaye Lunney of Esalen Institute, the idea of turning the trialogue into a public workshop emerged.

Ralph Abraham

My life has largely been ruled by the search for a certain iridescence, a certain glint or scintilla of noetic light that finds its way into a painting, a place, a book, or, in some few extraordinary cases, a person. Twenty years have passed since Ralph Abraham was brought to meet me by a friend from my high school days, who assured me that an extraordinary mingling of the minds would occur. This, in fact, has proven to be the case. Our separate journeys have taken us to many of the same places—to the Himalayas as well as the frontiers of mathematics and hands-on pharmacology.

Regarding Rupert Sheldrake, my travels and adventures in the Amazon had made me ever keen to explore the issues of theoretical biology. So naturally I jumped at the chance to meet him at a moment when awareness of the controversy surrounding his first book, A New Science of Life, *was just breaking over the American new-science crowd.*

For many years I've held wise and private consul with these two friends: Ralph, with steadfastness, humor, and insight; and Rupert, with qualities of kindness and gentleness that are rare in a scientific revolutionary. Only just before the conception of this book was it pointed out to me that a wider sharing of these discussions, dealing as they do with some of the great unfolding issues of all our lives, might offer valuable insight to others. Deep these trialogues may be, but the spirit in which they have been offered is one of three grown men deeply engaged in play. Our offer to the reader has been simply this: You can come along, too.

Terence McKenna

In 1981, a year before meeting Ralph and Terence, my first book, A New Science of Life, *was published in Britain. This was my attempt, as a biologist, to set out the hypothesis of morphic resonance, according to which there is an inherent memory in nature. The book had stirred up a great deal of controversy in Europe, especially when the international scientific journal* Nature *condemned it in an editorial entitled "A Book for Burning?" Having spent ten years doing research in one of the citadels of scientific orthodoxy, the biochemistry department at the University of Cambridge, I was well aware that the idea of collective memory, transmitted by a new kind of nonmaterial resonance, was not likely to win immediate acceptance. Having spent some six years in India, where I worked in an international agricultural research institute, I was also well aware that the mechanistic worldview of scientific orthodoxy was only one way of looking at the world.*

The book was published in the United States by Jeremy Tarcher in Los Angeles, and this brought me to California in 1982, my first journey to that frontier of the West; I had just turned forty. Early one morning Dan Drasin, a friend in San Francisco, put me on a bus to Santa Rosa, telling me he'd arranged a meeting there he felt sure I'd appreciate. I was to be met by someone called Terence McKenna. Sure enough, a puckish figure in dark glasses appeared at the bus stop in an aging Cadillac and took me off to his family home in the depths of Sonoma County. There I met his friend Ralph Abraham.

In England, especially at Cambridge, I enjoyed the discipline of mind imposed by the critical method, the historical awareness, the quickness of response, the active intelligence. In excess, however, it was oppressive. New ideas were treated as guilty until proven innocent, and as soon as I or anyone took off on a flight of speculation, the others opened fire. Shooting people down is a favorite sport of academics, and Cambridge is a free-fire zone.

In California, I found a sense of freedom from the past and an invigorating enthusiasm for the new, but much of it was shallow rooted and there was little place for wit, until I met Ralph and Terence. We had all been to India and had been much influenced by the unimaginable variety and complexity of cultural forms, the human warmth, and the speculative anarchy. We shared an interest in science and in the realms of consciousness beyond.

In the years since 1982 we've spent much time together trying out new ideas, developing old ones, and enjoying each others' company. Our conversations have a range and freedom I have rarely encountered elsewhere, and they have been for me the source of many new insights, as well as of inspiration and renewal. I hope that this book will encourage others to explore with their friends some of the questions we discuss here and will serve as a reminder of the importance of dialogue as a means of discovery.

Rupert Sheldrake

This book grew out of a series of discussions that had gone on for more than eight years. In our first discussion in 1982, we were delighted to find that our different areas of interest and styles of thinking were synergistic. Since that time, we have been able to explore our current interests together in ways that have been exciting and stimulating for ourselves and for others. Our friendship has been a source of inspiration in our individual research and writing.

From September 8 to 12, 1989, we met at Esalen Institute, in Big Sur, California, for four days of talk. The first two days consisted of a public program of dialogues and trialogues; during the second two days, we met privately. We met there again from September 3 to 6, 1990. The edited transcripts of our discussions at Esalen form the basis of this book.

We are entitling this book *Trialogues at the Edge of the West* both because these trialogues took place at the geographical edge of the West and because, in a metaphorical sense, they represent the thinking that is now possible at the leading edge of our culture. Each of us is a pioneer in new areas of thought: Ralph Abraham in the new theory of chaos; Terence McKenna in the revaluation of psychedelic experience and shamanic traditions; and Rupert Sheldrake in the new understanding of nature in terms of the evolution of habits. Where are these developments leading? What relationship do they have to each other, to religion and visionary experience, to our understanding of cosmic evolution, to the resacralization of the Earth, to the current ecological and political crisis, and to the coming of a new millennium? These are some of the questions we discuss in these trialogues.

In this book, we also explore the nature of the dialogue and trialogue themselves. Ever since Plato, dialogues have been recognized as a uniquely effective way of exploring the realm of thought: they are the basis of the dialectical method. But insofar as the dialectic of two points of view can result in a synthesis, it presupposes a third point of view that includes the two starting positions. We have found that trialogues have a more harmonious dynamic than dialogues with only two people, partly because the synthesis implicit in a fruitful dialogue can be made explicit by the third person. There is a current revival of interest in the dialogue form (perhaps the best-known example being the series of discussions between Joseph Campbell and Bill Moyers), and we hope that our book will help further the appreciation of the dynamics of dialogues and trialogues.

This book begins with a series of three dialogues in which each of us talks to each of the others and the third person comments toward the end of the discussion. In these dialogues, and in the trialogue that follows them, we are investigating the relationships among creativity, chaos, and imagination, and the relationship of these three concepts to the soul of the world. Rupert takes the point of view of evolutionary creativity, Ralph of chaos, and Terence of imagination. These discussions are followed by six further trialogues, led by each of us in turn.

We know of no comparable book. We hope that the excitement we have experienced exploring ideas together will communicate itself to readers. We believe that our experience of talking together and our personal affinity have enabled us to range widely and have given a cohesion to our discussions. This book should appeal to anyone interested in contemporary developments in scientific thought, the emergence of green consciousness, visionary experience, a new relationship to nature, the revival of religion, and the future of the West. We hope that this book will stimulate readers to continue these discussions in their own minds and together with their friends.

TRIALOGUES AT THE EDGE OF THE WEST

*What I suggest is the existence of a kind of memory inherent in
each organism in what I call its morphogenetic or morphic field. As
time goes on, each type of organism forms a specific kind of cumulative collective memory. The regularities of nature are therefore
habitual. Things are as they are because they were as they were.
The universe is an evolving system of habits.*
    —Rupert Sheldrake

*For me, the key to unlocking what is going on with history, creativity,
and progressive processes of all sorts is to see the state of completion
at the end as a kind of higher-dimensional object that casts an
enormous and flickering shadow over the lower dimensions of
organization, of which this universe is one.*
    —Terence McKenna

*There is another level . . . which I am calling Chaos, or the Gaian
unconscious. This contains not form but the source of form,
the energy of form, the form of form, the material that form is
made of.*
    —Ralph Abraham

1

CREATIVITY AND THE IMAGINATION

RUPERT: There's a profound crisis in the scientific world at the moment that is going to change science as we know it. Two of the West's fundamental models of reality are in tremendous conflict. The existing worldview of science is an unstable combination of two great tectonic plates of theory that are crashing into each other. Where they meet, there are major theoretical earthquakes and disruptions and volcanos of speculation.

One of these theories says that there's an unchanging permanence underlying everything that we know, see, experience, and feel. In Newtonian physics, that permanence is seen as twofold. First of all, there's the permanence of the eternal mathematical laws of nature considered by Newton and Descartes to be ideas in the mind of God—God being a mathematician. The image of God as a kind of transcendent disembodied mathematician containing the mathematical laws of nature as eternal Ideas isa recurrently popular idea, at least among mathematicians. The other sort of permanence is in the atoms of matter in motion. All material objects are supposed to be permutations and combinations of these unchanging atoms. The movement they take part in is also permanent and constant.

These permanences are summed up in the principles of conservation of matter and energy: The total amount of matter is always the same, and so is the total amount of energy. Nothing really changes at the most fundamental level. Nor do the laws of nature change. This model of the eternal nature of nature has been the basis of physics and chemistry, and to a large extent it is still the basis of physical and chemical thinking.

The other theoretical viewpoint is the evolutionary one, which comes to us from the Judeo-Christian part of our cultural heritage. According to the biblical account, there is a process in history of progressive development, but this process is confined to the human realm. In the seventeenth century, this religious faith was secularized in the notion of human progress through science and technology, and by the end of the eighteenth century the idea of human progress was a dominant idea in Europe. In the nineteenth century, through the theory of biological evolution, human evolutionary development came to be seen as part of the progressive evolution of all life.

Only in the 1960s did physicists finally abandon their eternal or static cosmology and come to an evolutionary conception of the universe. With the Big Bang theory, the universe became essentially evolutionary. This very recent revolution in science totally changed our worldview because the most fundamental thing in science is its cosmology, its basic model of the cosmos.

However, if all of nature is evolving, then what about the eternal laws of nature that scientists have taken for granted for so many centuries? Where were they before the Big Bang? There was nowhere for them to be, because there was no universe. If the laws of nature were all there before the Big Bang, then they must be nonphysical, idealike entities dwelling in some kind of permanent mathematical mind, be it the mind of God or the Cosmic Mind or just the mind of a disembodied mathematician. This assumption is something that physicists and most modern cosmologists have not yet begun to question seriously. It's an idea that's hanging over a theoretical abyss because there's no compelling reason to assume the laws of nature are permanent in an evolving universe. If the universe is evolving, then the laws of nature may be evolving as well. In fact, the very idea of the *laws* of nature may not be appropriate. It may be better to think of the evolving *habits* of nature.

The Big Bang theory is like the ancient mythological idea of the cosmos beginning through the cracking of a cosmic egg and continuing through the growth of the organism that comes out of it. This embryological metaphor is a developmental model. It replaces the notion of an eternal machine slowly running out of steam with the concept of a growing, developing organism that differentiates within itself, creating new forms and patterns. On Earth, this evolutionary process gives rise to all forms of microbial, animal, and plant life, as well as to the many and varied forms of human culture.

So how does this process happen? In my books *A New Science of Life* and *The Presence of the Past*, I attempt to explain how the habits of nature can evolve. What I suggest is the existence of a kind of memory inherent in each organism in what I call its morphogenetic or morphic field. As time goes on, each type of organism forms a specific kind of cumulative collective memory. The regularities of nature are therefore habitual. Things are as they are because they were as they were. The universe is an evolving system of habits.

For example, when a crystal crystallizes, the form it takes depends on the way similar crystals were formed in the past. In the realm of animal behavior, if rats are trained to do something in San Francisco, for example, then rats of that breed all over the world should consequently be able to do the same activity more easily through an invisible influence. There's already evidence, summarized in my books, that these effects actually occur. This hypothesis also suggests that in human learning we all benefit from what other people have previously learned through a kind of collective human memory. This is an idea very like that of Jung's collective unconscious.

Obviously, this is only part of the story. If the universe is a system of habits, then how do new patterns come into being in the first place? What is the basis of creativity? Evolution, like our own lives, must involve an interplay of habit and creativity. A theory of evolutionary habit demands a theory of evolutionary creativity. What gives rise to new ideas, to Beethoven's symphonies, to creative theories in science, to new works of art, to new forms of culture, to new instincts in birds and animals, to the evolving forms of plants and flowers and leaves, to new kinds of crystals, and to all the evolving forms of galactic, stellar, and planetary organization? What kind of creativity could underlie all these processes?

There seem to be two basic answers on the market. One is the materialistic viewpoint, which says that the whole thing is due to blind chance—that there are nothing but blind material processes going on, and then, by chance, new things happen. This viewpoint basically says, "There's no reason behind it. There's nothing intelligible about it. Creativity just happens."

The other theory is derived from the tradition of Platonic theology. It says that everything new that happens and every new form that appears corresponds to an eternal archetype, an eternal Idea in the mind of God, or an eternal formula in the mathematical mind of the cosmos.

Evolutionary creativity, however, is creativity that keeps on happening. It goes on as the world goes on. It's not something that just happened once in an act of creation at the beginning of the universe. Another model for understanding creativity is provided by our own imaginations, which are not full of fixed Platonic Ideas, but ideas that are ongoing and changing with a creative richness that continually surprises us.

Could there be a kind of imagination working in nature that is similar to our own imaginations? Could our own imaginations be just one conscious aspect of an imagination working through the whole natural world—perhaps unconsciously as it works underneath the surface of our dreams, perhaps sometimes consciously? Could such an ongoing imagination be the basis of evolutionary creativity in all of nature, just as it is in the human realm?

These are questions I want to put to you, Terence, because you've studied the realm of the imagination more than most of us.

TERENCE: Well, certainly I think that the relationship between creativity and imagination is the place to focus if we want to understand the emergence of form out of disorder. The whole notion of "eternal" laws of nature comes under question in the face of the Big Bang. Where were these laws before the Big Bang? One either has to hypothesize a kind of Platonic superspace in which, for reasons unknowable, these laws were present or say that somehow the laws of nature came into being complete and entire at the moment of the Big Bang. It's very hard to see how complex laws of nature such as gene segregation could exist in the situation of high-temperature physics and nonmolecular systems that prevailed at the beginning of the universe. In my thinking about how patterns came to be in the universe, I've attempted to take all the orthodox positions and stand them on their heads. I think it's a useful way to begin.

Is it credible that perhaps the universe is a kind of system in which more advanced forms of order actually influence previous states of organization? This is what emerges in Ralph Abraham's work with chaotic attractors. These attractors exert influence on less organized states and pull them toward some kind of end state.

For me, the key to unlocking what is going on with history, creativity, and progressive processes of all sorts is to see the state of completion at the end as a kind of higher-dimensional object that casts an enormous and flickering shadow over the lower dimensions of organization, of which this universe is one. For instance, in the human domain, history is an endless round of anticipation: "The Golden Age is coming." "The Messiah is immediately around the corner." "Great change is soon upon us." All of these intimations of change suggest a transcendental object that is the great attractor in many, many dimensions, throwing out images of itself that filter down through lower dimensional matrixes. These shadow images are the basis of nature's appetite for greater expression of form, the human soul's appetite for greater immersion in beauty, and human history's appetite for greater expression of complexity.

When I think about the terms *chaos, creativity,* and *imagination,* I see them as a three-stroke engine of some sort. Each impels and runs the other and sets up a reinforcing cycle that stabilizes organisms and conserves processes caught up in the phenomenon of being. This is a self-synergizing engine whose power emerges out of chaos, moves through creativity, travels into the imagination, returns back into chaos, then extends out into creativity, and so on. It operates on many levels simultaneously. The planet is undergoing a destiny.

The model we all take for granted—deep time, the time of geology—was only discovered in the nineteenth century. It's cosmically ennobling to think of the universe as a thing of great age, but I think it's time to put in place, next to the notion of deep cosmic time, the notion of chaotic sudden change, unexpected flux, sudden perturbation. As we've pushed our understanding of the career of organic life back nearly three billion years, the study of deep time has revealed tremendous punctuation built into the universe. As an example, recall the asteroid impact that happened sixty-five million years ago from which nothing on this planet larger than a chicken walked away.

The message of deep time is: we may not have as much time as we thought; the universe is dynamic, capable of turning sudden corners. This situation demands a new attitudinal response in which imagination is a kind of beacon—a scout sent ahead that precedes us into history. Imagination is a kind of eschatological object shedding influence throughout the temporal dimension and throughout the morpho-genetic field.

If the morphogenetic field is not subject to the inverse square laws that indicate decreased influence over distance, then I can't see why it couldn't be located at the conclusion of a cosmological process. One of the things that's always puzzled me about the *Big Bang* is the notion of singularity. This theory cannot predict behavior outside its domain, yet everything that happens and all our other theories follow from it. The immense improbability that modern science rests on, but cares not to discuss, is the belief that the universe sprang from nothing in a single moment. If you can believe that, then it's very hard to see what you can't believe. Such intellectual contortions are commonplace in science in order to save particular theories.

I propose a different idea that I think is eminently reasonable: As the complexity of a system increases, so does the likelihood of its generating a singularity or an unpredictable perturbation. I imagine the preexistent state of the universe to have been extremely simple, perhaps an unflawed nothingness. This is the least likely situation in which you'd expect a singularity to emerge.

If we look at the other end of the historical continuum, at the world we're living in, we see tremendously complicated, integrated, multileveled, dynamic complexity. And with every passing moment, the world becomes more complex. There are 106 elements. There are tremendous gradients of energy ranging from what's going on inside pulsars and quasars to what's going on inside viruses and cells. There are tremendous organizational capacities at the atomic level, the level of molecular polymerization, the level of membranes and gels, the level of cells and organisms, the level of societies, and so forth.

Evolution, history, compression of time—all these things are indications of the increasing complexity of reality. Is it not reasonable to suspect that a singularity must emerge near the end of the complexification process, rather than at its beginning? When we reverse our preconceptions about the flow of cause and effect, we get a great attractor that pulls all organization and structure toward itself over several billion years. As the object of its attraction grows closer to its proximity, the two somehow interpenetrate, setting up standing wave patterns of interference in which new properties become emergent, and the thing complexifies. To my mind, this is the Divine Imagination, as Blake called it.

Rupert and I were chatting last night in our room about the aboriginal nature of God, the idea that somehow time is the theater of God's becoming. From the point of view of a higher-dimensional manifold, God's existence is a kind of *fait accompli*. This is a paradox but not necessarily a contradiction, because in these ontologically primary realms we must avoid closure and hold on to the notion of a *coincidencia oppositorum*, a union of opposites. A thing is both what it is and what it is not, and yet it somehow escapes contradiction. That's how the open system is maintained. That's how the miracle of life and mind is possible.

I think of the Divine Imagination as the class of all things both possible and beautiful in a kind of reverse Platonism. The attractor is at the bottom of a very deep well into which all phenomena are cascading and being brought into a kind of compressed state. This is happening in the biological realm through the career of the evolution of life. It's simultaneously happening in the world as we experience it within our culture, in what we call history. History is the track in the snow left by creativity wandering in the Divine Imagination. In the history departments of modern universities, it is taught that this track in the snow is going nowhere. The technical term is *trendlessly fluctuating*. We're told that history is a trendlessly fluctuating process: it goes here, it goes there. We just wander around. It's called a random walk in information theory.

This is all very interesting, for we've begun to see, through the marvel of the new mathematics, that random walks are not random at all—that a sufficiently long random walk becomes a fractal structure of extraordinary depth and beauty. Chaos is not something that degrades information and is somehow the enemy of order, but rather it is something that is the birthplace of order.

RUPERT: Your description of the imagination emanating from the cosmic attractor sounds to me like a combination of Plato, Thomas Taylor, and Teilhard de Chardin. It resembles the Omega Point that, according to Teilhard, is the attractor of the whole evolutionary process. Everything is

being drawn toward this end point. This is like Aristotle's conception of
God as the prime mover of the revolving heavenly spheres. According to
him, the heavens were not being pushed by God, they were being pulled
by God. God is so attractive that the heavens keep on going round and
round, eternal rotation being the closest they can come to the divine state
of eternal bliss.

This idea of attraction has ancient roots, and Teilhard's is an evolutionary version of it. I agree with him and with you, Terence, that we
need the notion of an attractor to understand the evolutionary process of
the cosmos. This is the subject of contemporary discussion in the context
of the Anthropic Cosmological Principle, the idea that the cosmos must
have been constituted so that it allowed the evolution of carbon-based
life on at least one planet and then allowed the evolution of human
intelligence so that it could give rise to human cosmologists. Other
people are a kind of by-product.

If there is an attractor in the evolutionary process, which I think
there must be, then the question is, How does it work in the process of
evolutionary creativity? It could be a kind of future Platonic mind, which
is what you, Terence, seem to suggest. This Cosmic Mind contains all
possible forms and archetypes that are way out there in the future, and it
somehow interacts with what's going on now.

The way I understand it, there is an ongoing system of habits in the
cosmos built up from the past. Habits have a certain density, and matter
is dense because it's so deeply habitual. There's a sense in which the sheer
materiality of the natural world, and its sheer resistance to the imagination, is due to the fact that everything is so deeply imbedded in habit.
Left to themselves, habits would just fossilize, and the whole world would
become intensely, repetitively habitual. But there are other processes
going on, such as cosmological expansion and the continued presence of
chaos within the universe. This means that habits are continually being
disrupted by accidents—for example, asteroids hitting the Earth. In
our own lives, habits are continually being disrupted by developmental
changes and unexpected accidents, creating vacuums in which new
conditions and new possibilities can happen.

As I understand you, you are saying that the needs, the problems,
the tensions, the ongoing crises of the present somehow interact with

the cosmic attractor at the end of time, and it's as if sparks pass between them. Situations or problems attract to themselves aspects of the divine mind that are appropriate to their present circumstances, creating a kind of imaginative penumbra around what's actually happening. Similarly, our own imaginations are stimulated by what we're interested in, and our dreams reflect our preoccupations and interests and hidden motivations. The evolutionary imagination works by a kind of spark between the divine mind, or cosmic attractor, and present situations open to creativity.

I'm suspicious of the idea that everything that can possibly happen already exists somewhere. This concept denies creativity, reducing it to the manifestation of a future potentiality or possibility that is at the same time eternal. The final unified attractor and the primal unified state of the Big Bang have a symmetrical relationship. They're both part of the familiar model of history in which the end in some sense reflects the beginning, or the end in some sense is the beginning of a higher turn of the spiral.

I'm interested in the possibility that the imagination isn't all there, all worked out in potential form in advance—the possibility that the world truly is made up as it goes along. Henri Bergson, in his book *Creative Evolution*, strongly emphasizes that evolution implies ongoing creativity but that human beings will do anything we can to avoid this notion because it's so extremely difficult to conceive of ongoing creativity. I agree with him. We either have the tendency to reject the question by saying ongoing creativity is entirely random so we can't think about it, or else we substitute some sort of Platonic realm where everything is already present as an archetype.

I'm trying to look at a third possibility in which the imagination, instead of emerging from the light of the future or from a kind of Platonic mind, may emerge from something more like the unconscious mind—coming into the light from darkness. The formative process of the imagination may not be like sparks leaping from the mind of God but like new forms welling up from the womb of chaos.

TERENCE: It seems to me the problem revolves around the notion of purpose. Specifically: Is there one? If there is one, what is it? Nineteenth-century science went to great lengths to eliminate purpose from all of its model building in order to make, once and for all, a clean break from the contaminating power of deism. For instance, in evolutionary theory the great breakthrough in the nineteenth century was the notion of random process. Not knowing that it was background cosmic radiation that drove mutation, scientists played two random processes against each other: what they called "sporting," or the production of variant types, and selection based on fitness to the environment. When these two random processes are run into each other, an exquisite order emerges—of animals, plants, and ecosystems. Darwinists could say, "You see, we have no need for God or purposes or divine plans. We show that out of the chaos of the moment emerges order."

This tendency was so strong in nineteenth- and early twentieth-century thought that evolutionary biology sought to entirely appropriate the word evolution. It was not to be used in any other context. A biologist once said to me, "If it doesn't involve genes, it isn't evolution." You couldn't talk about the evolution of the novel, the sonata, or socialism unless it involved genes. Largely through the work of Erich Jantsch, in his books *Design for Evolution, Evolution and Consciousness* (with C.H. Waddington) and *The Self-Organizing Universe*, this kind of narrow thinking was overthrown.

I don't believe that everything is finished in some deterministic sense at the end of the cosmos. I do believe that there is some kind of intimation of purpose that directs the evolution of processes through time and keeps them from simply becoming random walks. If we believe that all of the imagination is being made up in the present, we're back with the trendlessly fluctuating theorists of history. If none of it exists in the future, then there is no compass point by which to guide the process forward.

C.H. Waddington's idea of the chreode, described in his book *The Strategy of the Genes*, allows me to preserve your intuitive concept, Rupert, that everything is being made up as it goes along, as well as my own strong intuitive hit about the necessity for a vector point in the future. This is done by saying that time is a topological manifold over which events must flow subject to the constraints of the manifold. I call the surface of this manifold *novelty*, and I believe that by examining time

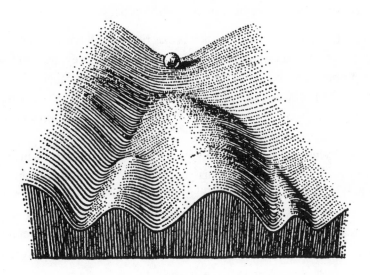

Figure 1. A Chreode. *Part of an epigenetic landscape, illustrating C.H. Wadding-ton's concept of chreodes as "canalized pathways of change." The chreodes corre-spond to the valleys and lead to particular developmental endpoints, which could, for example, be the sepals, petals, stamens, and pistils in a flower.*

from this point of view we can see when in history great outbreaks of novelty occurred. What's important for this argument is that—without knowing any of its content—we can place the novelty of novelties, the novelty to the nth power of novelty, at the end of the historical process and watch it operate as an attractor without having any information concerning its particulars. This point of view comes very close to Neoplatonism. We have to maintain the unknowability of God, hence the ultimate unknowability of the imagination. Nevertheless, we have to grant it as an attractor.

RUPERT: It's partly a question, as I see it, of what role one thinks the attractor has. One could say, as I think Teilhard or Aristotle conceived it, that the entire cosmic process is drawn toward states of higher unity. These states are not just higher states in general but as many possible states as can be, which explains why there are so many forms of life. The question is, are the new forms arising in the attractor, or is the attractor simply attracting what's already a diversity of forms through a process that lies between them, which would be the process of imagination?

TERENCE: As I see it, the attractor is exerting an influence and pulling these things toward it, but not in a direct trajectory. They must follow this topological manifold of time; they have to enter the labyrinth, the interstices of becoming in order to reach the attractor. That's why history is so perverse. Two steps forward, one step back, is still a prescription for a kind of progress.

The Divine Imagination into which history takes us, and which our tools and cultural adaptation clarify and define for us, is something we have been moving toward rather like someone driving using only the rear-view mirror. What is exactly dead ahead of us can hardly be seen. Historical compression increases daily throughout global culture until there is almost a sense of free fall into the deeper fields of the attractor.

RUPERT: I think we should ask Ralph.

TERENCE: Yes, let's get Ralph in here.

RALPH: I'm trying to see whether imagination and creation are the same or different, and whether you're having a disagreement or are describing the same thing in different ways.

The idea of the attractor is related to the problem of the source of new forms. Rupert suggests that new forms come out of chaos, or from the unconscious. There's still the geometric question, Where are they coming from? Your disagreement, as I see it, is in the location of this source of new forms. Are they projected from an attractor at the end of time, or do they come out of the dynamics working within the field in the evolution and habits of nature? In this geometric model of the world's soul, your only difference, I think, is in perspective. Or perhaps in your idea of time. Behind us is history and in front of us is possibility. In the determination of the next moment, your conflict is just a difference of metaphor, not of process. You seem to agree that the next moment is created out of the present moment through a process involving creativity, imagination, chaos, and a world of possibilities located somewhere.

TERENCE: It occurs to me that creativity might be a lower-dimensional slice of the Divine Imagination, a process that seeks to approach this thing that somehow has an all-at-once completeness about it. Does that fit with your notion of creativity as movement toward the realization of a kind of ideal realm?

RUPERT: I think creativity seems to involve a process like the welling up or boiling up of new forms in an incredible diversity. New forms are conditioned by memories of what has gone before and by existing habits, but they are new syntheses, new patterns. There could be a kind of unifying process at work such that anything that emerges above the surface of the unconscious or the darkness or the chaos has to take on a kind of wholeness to come above that surface. It has to take on a unified form. But it could be any unified form.

One model for this creative process is dreaming. Dreams involve the appearance of stories and symbols and images that we don't create with our conscious minds. In fact, we usually just forget this whole wonderful display of psychic creativity that happens for each of us nightly. When we remember our dreams, they're bizarre and unexpected. It seems almost impossible to have an expected dream. This curious feature raises the question of where dreams come from. The Jungians would say that they come from structures and processes in the darkness of the collective unconscious. They'd see them not as descending from some higher world but as welling up.

The human imagination obviously works through dreams. It works through language, through conversations, through fantasies, through novels, through visions and inspirations, and it is also revealed through psychedelics in a particularly extreme form. In what sense is this imagination that we know from our own experience related to the imaginative creative principle of nature? Is there a kind of Gaian dreaming? For example, is the Earth, Gaia, awake on the side of it that's in the sunlight? As it rotates, is the side that's in the darkness dreaming? At night, are plants, animals, and whole ecosystems in some sense in a dream state, when dreams and spontaneous images of what might be possible come to them? What form would a Gaian dream take? Or what form would a Gaian psychedelic experience take?

TERENCE: I think a Gaian dream would be human history. Perhaps the planet's been sleeping for fifty thousand years and is having a dyspeptic dream that causes it to toss and turn. If it could only awaken from that dream, it would just shake its head and say, "My God, I don't know what it was, but I hope it doesn't come back!" Human history has that quality. James Joyce, in his book *Ulysses*, has Stephen Daedalus say, "History is the nightmare from which I'm trying to awaken." The whole structure of Joyce's novels involves the integration of historical data with daily newspapers and that sort of thing to evoke the quality of a dream.

In mentioning psychedelics, Rupert, you said they revealed an extreme and intense example of this upwelling creativity. I think what they reveal is so intense and extreme an example that it argues strongly that the imagination is not the human imagination at all. While we may be able to analyze dreams and see the acting out of wish fulfillment or repressed sexual drives or whatever, depending on our theory of dreams, the psychedelic experience at its intense levels goes beyond the terms of human motivation. It seems rather to enter an ontological reality of its own, one that the human being is simply privileged to observe briefly. A deep psychedelic experience says no more about a person's personality than does the continent of Africa. They are, in fact, independent objects.

To my mind, the Divine Imagination is the source of all creativity in our dreams, in our psychedelic experiences, in the jungles, in the currents of the ocean, and in the organization of protozoan and microbial life. Wherever there is large-scale integration rather than simply the laws of physics, the creative principle may be beheld.

RUPERT: Do you think, then, that in psychedelic experiences you're actually tapping into the diurnal cosmic imagination?

TERENCE: Absolutely. Psychedelic experiences and dreams are chemical cousins; they are only different in degree. This is how I can see human history as a Gaian dream, because I think every night when we descend into dreams, we are potentially open to receiving Gaian corrective tuning of our life state. The whole thing is an enzyme-driven process. We are like an organ of Gaia that binds and releases energy. A liver cell doesn't need to understand why it binds and releases enzymes. Similarly, as humans, we bind and release energy for reasons perhaps never to be clear to us but which place us firmly within the context of the Gaian mind.

We have been chosen, just as indole acetic acid has been chosen in plant metabolism, to play certain roles. We have a role, but our role seems to be a major one. We are like a triggering system. Out of the general background of evolutionary processes mediated first by incoming radiation to the surface of the Earth and then by natural selection, suddenly we arrive, with our epigenetic capability to write books, tell stories, sing, carve, and paint. These are not genetic processes, these are epigenetic processes. Writing, language, and art bind information and express the Gaian mind very well.

I see each of us as a cell in communication with the Divine Imagination, which is sending images back into the past to try and direct us away from areas of instability. The Gaian mind is a real mind; its messages are real messages, and our task—through discipline, dreams, psychedelics, attention to detail, whatever we have going—is to try and extract its messages and eliminate our own interference so that we can see the face of the Other and respond to what it wants.

This isn't for me a philosophical problem. It's a problem that relates to the politics and action that we take as a collectivity and as individuals.

RUPERT: The idea that we tune in through our own imaginations to the Gaian mind seems attractive, and I think it fits quite well with dreams, psychedelic experience, imagination, and so on. The next question for me is this: How is the Gaian imagination related to the imagination of the solar system, and that of the solar system to the imagination of the galaxy, and that of the galaxy to the imagination of the cosmos, and that of the cosmos related to what we could call the imagination of the cosmic attractor, or the Omega Point, or the Cosmic Christ?

TERENCE: I'm not sure I want to follow you into the Cosmic Christ. I think there should always be some physical stuff to hang these things on. The Gaian mind is not a problem—the Earth teems with life. A Jovian mind is not a problem because the complex chemistry and metallic behavior of gas and ice under pressure seem to place enough cards on the table that mind could well emerge in that situation. Similarly, the oceans of Europa might be a friendly environment for life and mind. There are a number of places in the solar system where there's enough complex chemistry that I can imagine that sustained, self-reflecting processes might get going over billions of years.

However, to move from that level of mind to the hypothesis of a continuous hierarchy of minds, out to the level of the galactic mind, you have to ask hard questions. How long does it take the galactic mind to think a thought? Does it do it instantaneously via morphogenetic fields? If so, then what are the transducing and signal-sorting filters through which the thought travels? If through light, then to say that the galactic mind's thoughts are "vaster than empires and more slow" is to suggest that they are very high-speed phenomena indeed. Empires would come and go by the hundreds before a galactic thought could reach from one side of itself to the other.

RUPERT: We don't know enough to begin to answer these questions. I think a factor that changes everything is the discovery of dark matter— the fact that 90 to 99 percent of the matter in the universe is utterly unknown to us. This recent discovery effectively tells us that the cosmos has a kind of unconscious, a dark realm that conditions the formation and shapes of the galaxies, their interactions, and everything that's going on within them. Your search for the basis of the imagination in the known phenomena of physics is certainly an important one, but physics itself has revealed that there's so much more, and this dark matter could be the basis of any number of processes unknown to us.

TERENCE: I assume that psychedelics somehow change our channel from the evolutionarily important channel giving traffic, weather, and stock market reports to the one playing the classical music of an alien civilization. In other words, we tend to tune to the channel that has a big payback in the immediate world. It seems obvious to me that there are channels of the imagination that are not so tailored for human consumption. I think you're correct, that memories and hence all objects of cognition are not in the wetware of the brain. They are somehow plucked out of a superspace of some sort via very subtle quantum mechanical transductions that go on at the molecular level. The Divine Imagination is the reality behind appearance. Appearance is simply the local slice of the Divine Imagination.

RALPH: In figure 2, I'm presenting a personal map of the preceding dialogue on creativity and imagination so that we can locate chaos on the map. Let's assume there was a beginning and there will be an end. In the

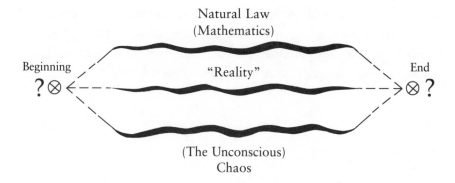

Figure 2. Three Levels of History. *As time advances to the right, three layers of history spring up from a point, The Beginning, and collapse again to a point, The End.*

meantime there is history, the time wave. There is a present moment also, so there is a past and a future, and the window of the present is moving along the time wave. Along the way, the future is being created by the emergence of forms of increasing complexity (according to Rupert) and increasing integrity (according to Terence).

There is another level below, at the bottom of the figure, which I am calling chaos, or the Gaian unconscious. This contains not form but the source of form, the energy of form, the form of form, the material that form is made of. Some little tinkle comes along like a ringing bell, then a form pops up and becomes part of the time wave.

There was also a question that arose briefly in the discussion, before being rejected forever, on the role of mathematics and natural law. This belongs to another level of reality, which I have drawn over ordinary reality in figure 2.

I think there are at least three different levels: ordinary reality, an undercurrent of chaos, and an overworld of law and order. This is not everything, not yet the full vision that any one of us has had in our own explorations of the larger space. These are simply some of the components of a larger thing that is the world soul, the all and everything.

There are some difficult questions. Do these other levels really have a beginning and an end, as this discussion has assumed ordinary reality does? Did natural law exist before the Big Bang, in which matter and energy were created? I putdotted lines and question marks in figure 2 to suggest these questions, which can be addressed later.

There's a cosmic imagination, the imagination of the anima mundi, the soul of the universe. Within this are the imaginations of galaxies, solar systems, planets, ecosystems, societies, individual organisms, organs, tissues, and so on.
    —Rupert Sheldrake

The more complex a structure, the more difficult it is to embrace with our minds. Words are frequently inadequate. Language has evolved through the necessity of sharing our experiences on a level of complexity that is more or less traditional and that is inadequate to understand the whole world, or the world soul, or the biosphere of planet Earth. Mathematics has only a little more magic than ordinary language.
    —Ralph Abraham

The modeling challenge for the future is human history. We will no longer be playing little games to demonstrate something to a group of students or colleagues, but we will actually be proposing models and methods powerful enough to begin to model the real world.
    —Terence McKenna

2

CREATIVITY AND CHAOS

RALPH: My own role here as chaos advocate is to encourage the fantasy that form arises from chaos. Chaos theory involves three levels, all of which are aspects of the present rather than of the whole of history as in figure 2.

In figure 3, the mathematical level is on top. This is the level or space of models, things that we do in our minds that create metaphors, images for other things. Ordinary reality is at the bottom level, and includes the matter and energy world as well as the mental world. This level might include our bodies and minds, the bodies and minds of microbes, and the Gaian body and mind. These two levels have supported the history of the sciences since Newton, or maybe even since Pythagoras, and may correspond to the upper two levels of the tableau of the preceding dialogue (figure 2).

Recently, a new level has appeared, interpolated between these two levels. This is a very interesting half step, created by the computer revolution and the development of computational mathematics around 1960. This interpolated level makes our discussion a little more confusing because the word model now means either a mathematical model—for example, an ordinary differential equation—or a computer program that simulates the mathematical model. The computer model is more real than the mathematical model but less real than ordinary reality.

In this modeling context, there are models for chaotic behavior called *chaotic attractors* and models for radical transformations of behavior called *bifurcations*. From chaotic attractors and their bifurcations, which live on the math level, we gain experience and get a feeling as to the appropriateness of a question that might live in the historical realm (figure 2)—which is somehow more real than the metaphorical realm (figure 3). Whether or not, in the Gaian mind, form can be pulled out of chaos by the ringing of a bell or something, we can only speculate.

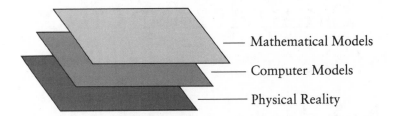

Figure 3. Three Levels of Reality. *Three levels of reality at a given moment in time. The mathematical universe is shown above the level of ordinary (consensual) reality of everyday life, matter, and energy, as in the ancient model of Plato. But sandwiched in between is a new level, created by the computer revolution.*

However, there are experiments with mathematical models for chaos that might be relevant to this question.

For example, a good laboratory for the study of chaotic dynamics is the dripping faucet. The dripping faucet was discovered as an ideal demonstrator for chaos theory because lectures are usually given in a physics lecture hall and they always seem to have sinks and faucets in the front. When you crack the tap a little bit, the water drips out very regularly. If you crack the tap a little more, the drips speed up, but they're still regular. When you crack it a little more, they sound irregular, like rain dripping off a roof. If you measure the time between drops, and make a list of these numbers, you have the paradigmatic example of a chaotic time series.

Somebody decided to study this dripping faucet seriously after seeing it in physics lectures. This person, Rob Shaw, is now one of the leading people in chaos theory. He did a very fine study by placing a microphone in the sink where the drop would hit it, getting an electronic beep, putting the time intervals into his computer, and analyzing the results. These portions of his study all belonged to level one, the lowest layer of figure 3, the physical world. Then he made a model for the dripping faucet on level three. In this mathematical model, the water drop gets bigger and bigger, and when its mass reaches a critical value the drop falls off the faucet. From this model on level three, Shaw wrote a computer program to simulate it, which was a model on level two. He ran the simulator and produced data that was almost exactly like the experimental data from the actual faucet on level one. Opening the model's tap eventually changed the simulated data from periodic to chaotic through a bifurcation.

This is an example of modeling in the three-level context. The point of this kind of modeling is to gain understanding as part of a hermeneutic circle. You look at the data, try to build a model, and you fail. You then observe in a different way, which helps you to build a better model, and, as the circle turns, the level of your understanding grows. This is what Rob Shaw did with the dripping faucet. The different way of observing data that came to him from the model was a method now known as *chaoscopy*. In this method, you take the sequence of numbers— the time between drops—and visualize a vertical column of numbers. Then you make a copy of this column of numbers over to the right. You whack one number off the top of this second column and move the entire column up one number. Now you have a column of pairs of numbers. Then you plot these pairs in a plane figure as a series of points.

There's a film available from Aerial Press that shows a machine actually doing this. From totally chaotic data viewed in this particular way, through chaoscopy, you get a set of points in a plane. If the data were really random, the dots would be all over the plane. Instead they lie along a smooth curve! This indicates a chaotic attractor.

A hidden order in chaos is revealed by a new way of looking. From the observation of the data in this way, the smoothness of the curve suggests, to a chaos theorist, a model that you can actually take off the shelf and apply to other data. There are models on level three that are good for understanding certain behavior on level one. On level two, an intermediary can either create the mathematical model on level three from the real data of level one, or create simulated data from the mathematical model to compare with the real data.

I have, I hope, arrived at an actual connection between chaos theory and the discourse we're trying to carry on here to increase our understanding of our past, our future, and our possibility of even having a future.

RUPERT: The problem I have with chaos theory is that I'm never quite sure what it's saying. There seem to me to be two things that are of interest in it. One is the actual detailed models that chaos theoreticians make. These theoreticians are finding fairly simple equations that will generate complicated and seemingly chaotic structures. The other aspect of interest is more general. Those who make mathematical models of chaos have given scientists permission to recognize that in fact there's an inherent indeterminacy throughout the physical world.

In the nineteenth century, it was generally believed there was no indeterminacy at all. Everything was believed to be totally determined by eternal laws of nature. Laplace thought that the whole future and past of the universe could be calculated from its present state if there were a mind powerful enough to do the calculations and make the observations. This illusion of total predictability held science under its spell for generations. Scientists were dazzled by its imaginary power. Of course, they couldn't calculate everything and they still can't. Far from predicting the entire future of everything, they still can't predict the weather very accurately a few days from now. The ideal of total predictability in principle was no more and no less than an act of faith.

With quantum mechanics, in 1927, came a recognition of genuine indeterminism in nature. Since then, there's been a gradual recognition that indeterminacy exists not only at the quantum level but at all levels of natural organization. There's an inherent spontaneity and indeterminism and probability in the weather, in the breaking of waves, in turbulent flow, in nervous systems, in living organisms, in biochemical cycles, and in a whole range of phenomena. Even the old-time favorite model for total rational mathematical order, the orbits of the planets in the solar system, turns out to be chaotic and unpredictable in terms of Newtonian physics. This indeterminism is now being recognized at all levels of nature.

It seems to me that this openness of nature, this indeterminism, this spontaneity, this freedom, is something that corresponds to the principle of chaos in its intuitive and mythological senses. Mathematicians have used the word *chaos* in a variety of technical senses, and it's not entirely clear to me how these technical models of chaotic systems correspond to intuitive notions of chaos.

What I'd like to consider, through a familiar physical process such as cooling, is the way in which form appears out of chaos. If you start with something at a very high temperature, atoms don't exist in it. Electrons fly off the nuclei and you get a plasma, which is sort of a soup of atomic nuclei and electrons, with its own distinctive properties. If you cool the plasma to a certain temperature, atoms begin to form. Electrons start circulating around nuclei, and you get a gas of atoms. But the temperature is still too high for any molecules to form. If you cool it down further, you get molecules. If you cool the system down even further, you get a stage at which more complex molecules come into being. Still, they're gaseous. Cool it down further, and they turn into a liquid that can form drops and flow around and have quite complex, ordered arrays of molecules within it. Cool it further, and you get a crystal that is a highly ordered, formal arrangement of atoms and molecules. You get a progressive increase in complexity of form as you lower the temperature.

In traditional kinetic theory, lowering temperature gives less random kinetic motion of particles; there is less chaos and an increase in complexity of form as the cooling process takes place. We all know about the cooling of steam into water and the cooling of water into ice crystals, as in snowflakes and frost. This formative process occurs as thermal chaos is reduced. The opposite happens if you warm things up. So there seems to be an inverse relationship between chaos and form.

In a sense, that's what's happened in the entire universe. We're told that the universe started off exceedingly hot—billions of billions of degrees centigrade. It was so hot that stable forms were not able to emerge. By expanding, it cooled down. When it was cool enough, atoms emerged, then stars and galaxies condensed, then solar systems and planets. Planets are the cooled remnants of exploding stars. The elements in us and in our planet are stardust, formed from supernovas. The evolutionary appearance of form comes about through cooling; form emerges progressively from chaos.

How do these new forms come into being? This is the big problem of evolutionary creativity. How did the first zinc atoms come into being? The first methane molecules, the first salt crystals, the first living cells, the first vertebrate? How did the first of anything come into being in this evolving universe?

One way of looking at this problem is to see the expansion and cooling process, and indeed the flow of events in general, in terms of the flux of energy. The concept of energy, which is one of the great unifying concepts of physics, was formulated in the nineteenth century. It's not entirely clear what energy is. In some sense it is the principle of change. The more energy there is, the more change that can be brought about. In this sense, it is a causative principle that exists in a process. This process, the energetic flux of the universe, underlies time, change, and becoming, and it seems to possess inherent indeterminism. The energetic flow is organized into forms by fields. Matter is now thought of as energy bound within fields—the quantum matter fields and the fields of molecules and so on.

I think there are many of these organizing fields that I call the morphic fields, and that they exist at all levels of complexity. These fields somehow organize the ongoing flux of energy that is always associated with chaotic qualities. Even organized systems of a high level of complexity, such as human brains, have this probabilistic quality. The fields that organize this energy giving rise to material and physical forms are themselves probabilistic. Chaos is never eliminated. There's always an indeterminism or spontaneity at all levels of organization.

There are two principles: a formative principle, which is the fields, and an energetic principle. Energy is the principle of change, and pure change would be chaos. One way of thinking of these two principles is in terms of the Indian Tantric notion of Shakti as energy and Shiva as the formative principle working together to create the world we know.

If the formative principle operates through the fields of nature, then how do these fields operate? How are these fields governed? How do they have the forms, shapes, and properties they do?

I think the organizing fields of organisms are what I call morphic fields, and that these fields contain an inherent memory. They are essentially habitual, and nature is the theater in which these habitual fields organize the indeterminate flux of energy. The fields themselves, by having this energy within them, share this indeterminate quality also.

This brings us to the question of creativity. How do new fields and new forms come into being in the first place? Where do they come from? This morning, Terence and I were discussing how they may arise out of the interaction of chaos and some kind of formative, unifying aspect of the cosmic mind, which you, Ralph, hijacked for the Pythagorean sect by calling it the realm of mathematics. There's an interaction between these two levels, which you've shown by the wiggly line in the middle of your diagram, indicating the world of becoming.

This brings us back to the nature of what you call the mathematical realm and what I call the formative realm. Is there a kind of mathematical realm for the universe, somehow beyond space and time altogether, which conditions all forms of creativity and all patterns and possible systems of organization that come into being in the world? Or are these all made up as the evolutionary process goes along? These are questions Terence and I touched on in our dialogue on creativity and the imagination [chapter 1].

I think if we take the view that things come into being as evolution goes along and that the cosmic soul has a kind of imagination, then we can think of form as coming into being through the imagination as nature goes along, and we can see this imagination as having many levels. There's a cosmic imagination, the imagination of the *anima mundi*, the soul of the universe. Within this are the imaginations of galaxies, solar systems, planets, ecosystems, societies, individual organisms, organs, tissues, and so on. There are many levels of organizing souls and imaginations. We don't have to leap straight from the level of a molecule or a plant cell to the level of Divine Imagination, or to the transcendent realm of mathematics. There's a whole series of imaginations in between.

My view is that there isn't a kind of mathematical mind out there, already fixed, already full. What people do is make mathematical models of various aspects of nature and then project these models on nature, creating the illusion that they're the real thing. The result is that the cosmic imagination seems to be engulfed within an eternal mathematical mind, when it may be no more mathematical than human beings are when we're dreaming. We do not experience our dreams as being generated by equations, or as essentially mathematical in nature.

What I'm suggesting is that the cosmic imagination might include within it a mathematical realm, and that this mathematical aspect is evolving just as our own understanding of mathematics is evolving in time.

RALPH: I don't see that mathematics is substantially different from verbal description as a strategy for making models. For example, I described a geometric, visual model for the all-and-everything, including within it the world soul and so on. If drawn as a geometric picture instead of a word picture, that's officially mathematics—that's geometry. I think that with mathematics we can make a model for anything. Mathematics can be regarded as simply an extension of language. It's not a code of laws describing the universe, although that is its usual paradigm.

Mathematics is a particularly good language for describing, discussing, and imagining things that are really complicated. The more complex a structure, the more difficult it is to embrace with our minds. Words are frequently inadequate. Language has evolved through the necessity of sharing our experiences on a level of complexity that is more or less traditional and that is inadequate to understand the whole world, or the world soul, or the biosphere of planet Earth. Mathematics has only a little more magic than ordinary language.

Rupert had a complaint about chaos theory, about determinism and prediction. According to chaos theory, prediction and determinism are impossible! Even though it uses the language that deterministic thinkers use, with respect to its technical details, there is only a sort of probabalistic prediction.

The models for the things you want to talk about, such as cooling, don't come from chaos theory alone. They come from bifurcation theory, which exemplifies the best of mathematics. Mathematics says that, based on particular assumptions, certain things will not happen and only certain other things will. You get a list of three or four of these so-called bifurcations. With cooling, for example, you may have a control knob with which you are turning down the heat under a pan. The boiling gradually subsides to simmering, which subsides to a little bit of waving, which subsides to nothing. At each stage, based on the mathematics, there is a model that has attractors, maybe chaotic attractors. Every time you change the knob, you get a different model. Therefore, if you can't predict how the knob is going to change, the models won't give any prediction at all. The only thing bifurcation theory can tell you is to expect certain transformations.

For example, Terence has pointed to the punctual aspect of evolution—the fact that many transformations are saltatory, catastrophic, abrupt. Bifurcation theory simply says that, in models of this type, most of the transformations are abrupt. It says that determinism, even probabilistic determinism, is impossible using these mathematical models. There is an encyclopedia of bifurcations that are very good models for transformations, for the emergence of form as in the Neolithic revolution or the formation of the solar system.

Bifurcation theory can be used to model everything, so it never settles the question of the origin of things or the true nature of ordinary reality. Like language, it's good for communication; it's good for understanding, because modeling is part of our basic process of understanding. Models are no good for prediction, but they're good for the growth of understanding.

RUPERT: You put the whole case impressively and modestly in claiming that mathematicians are just making models. Chaos theory provides a new range of models. We can look forward to more models coming onto the market in the future.

I suspect the traditional assumption is that successful models work because there are mathematical aspects of nature to which they mysteriously relate. I often meet mathematicians or physicists who say quantum physics is the most brilliant predictive system that mankind has ever known, predicting things to many places of decimals. When I was working in India as an agricultural scientist, there was no one who could predict the outcome of my crop experiments. Crops growing in fields were miles beyond the capabilities of any modeling process based on the fundamental principles of physics. One could produce empirical, string-and-sealing-wax models for crop production and run them on computers, but none of them provided to me a convincing demonstration that the whole thing depended on a hidden mathematical order.

Are the fields of nature more real than the mathematical models we make of them, or is there a kind of mathematics yet more fundamental than the fields? For example, take the polarities of the electromagnetic field: north and south magnetic poles, and positive and negative electric charges. Do these polarities exist because of some fundamental Two Principle, an archetypal duality behind and beyond nature? Or is that just the way fields are? When we look at a wide range of polar phenomena, we make an abstraction that we can then model mathematically, but it doesn't exist in some objective, transcendent realm.

RALPH: You could ask different mathematicians and get different answers. I'll give you mine, and other mathematicians will say it doesn't count because I'm *not* a mathematician, and my answer is the proof of that.

For me, mathematics is a beautiful landscape, an alternate reality, filled with possibilities not yet seen. There are some older and some younger parts in the mathematical landscape, and this entire system is in coevolution with ordinary reality as people enter and hang out there to study and invest their creative energy. In this mathematical landscape, there is only a small part that has been used for modeling anything in ordinary reality.

From the viewpoint of a nonmathematician, the part of mathematics that has been used for modeling something familiar, like the simmering of hot water, is the only part visible. Nonmathematicians may exclaim about the amazingly perfect fit between a mathematical concept evolved solely in the human mind and a boiling pot of water. This part of mathematics became visible primarily during the history of physics, which is devoted to the study of the simplest possible systems.

When you talk about your experience as an agricultural scientist, you are talking about a realm that is infinitely more complicated than the most complex physical system. The parts of mathematics that have been used by physicists are the parts that are the least interesting to mathematicians. Mathematics offers much more to the complex sciences than it offers to physics. The whole potential of mathematics to aid us in our evolution comes from the fact that it can extend our understanding of systems that are too complex to understand without it, as when a small change in the weather causes some peas to grow at the expense of others.

In an ecosystem, there are so many different things. We can't be sure that a single oil spill off a single coast could not produce desertification or bring on the equivalent of nuclear winter. Our understanding can be advanced by mathematics, because mathematics is an extension of language for dealing with complex systems. We can have models of emotional relationships, of love affairs, of arms races between nations, or the United Nations. We can model these things with models that are not perfect, but they're better than no models. The construction of these models is part of our evolution, and it's part of the evolution of the mathematical landscape as well.

RUPERT: I must admit that my interest in mathematical models has enormously increased since I came across attractors. No one in any other branch of science has been able to think in terms of teleological principles that pull from in front. You, Ralph, have done more than any mathematician I know to make the essential features of this kind of mathematics accessible. There's not a single equation in your four volumes on visual dynamics. Through diagrams, you give the essence of what dynamic systems are. Normally, mathematical ideas are hidden behind an opaque cloud of symbols that most of us can't penetrate. It's as if our only experience of music was looking at the scores of symphonies without ever actually hearing the symphonies themselves. These symbols refer to things that for real mathematicians are visual intuitions.

Attractors have really changed our way of thinking about nature because they've made it possible to think about what Aristotle called the entelechy, the end that attracted toward itself the process of change. What I'd like to know is how you think attractors work. No matter how we try to get out of it, they seem to imply a pulling from in front rather than a pushing from behind, something that is more Aristotelian than mechanistic. At the cosmological level, we arrive at what Terence and I were discussing this morning—the idea of an attractor for the entire cosmic evolutionary process.

RALPH: I have to admit that when I heard this from both of you my jaw dropped. I was astonished at this interpretation, and I can't say it's wrong, but I'm sure it's different from the way any mathematician has thought about attractors in dynamical systems before.

Imagine a train going down a track, and it's going to get to the next station in seven minutes. Is the station pulling the train? The dynamical system is the track, the present moment is the train, and the attractor is the station. This attractor might not be simply a point—it might be a circular section of track, or it might be a tangled heap of track, a chaotic attractor. The idea of the attractor pulling the train may be suggested by the word *attractor*. When we thought of this word in the early 1960s, we never thought it would be interpreted in this way. Now I can see that it's the obvious interpretation anyone would make when they read this word.

RUPERT: The same problem was confronted by Sir Isaac Newton, when he chose the word *attraction* for gravitation. When Voltaire visited London in 1730, more than forty years after Newton had published his ideas, these ideas were still not accepted in France for the principal reason that Newton had used the word *attraction*, with its connotations of sexual attraction and its animistic and subjective associations. The idea that the Earth could be attracting a stone like an attractive woman attracts a man seemed ridiculous. Voltaire said that if Newton had chosen a different word his theories would have been adopted thirty years earlier.

The truth is that mechanistic cosmology replaced animistic cosmology through introducing animistic principles such as gravitational "attraction" by subterfuge, pretending that they were mechanical principles. In Aristotle's view of the world, stones fell because they were attracted to the Earth; they were attracted to their proper place; they were going home. Newtonian physics said that it was completely wrong to think of nature working on the basis of such attractions. Instead, stones fell because of "gravitational attraction"! Isn't attraction a strange concept to use if one is trying to deny attraction? The Newtonian tried to forget about animistic associations and pretended that this was just a neutral technical term.

In the evolution of science since the mechanistic revolution, attraction has been reinvented again and again. I suspect that the same is true, Ralph, of your dynamical attractor. Choosing the words *attraction* or *attractor* gives the idea an inherent appeal and plausibility.

I think that even if you take your example of the train, the station may not be pulling the train but there's a sense in which the station really is an attractor. I get on a train that's going to London because it's my purpose to go to London. In a sense, what's motivating the train is the purpose of the people getting to London or New York or Los Angeles. Unless human beings are purposive and have destinations they want to get to, and unless railway companies have schedules and plan the way they run the trains in accordance with supply and demand, the train won't run. The train can be modeled as if it's just a dynamical system running along rails that happen to end up in London. If you observe enough trains on the London railway line, you see lots of them going to London, so you put London in the model as an attractor, but at the same time you say the destination has nothing to do with attraction. This is, in my opinion, a subterfuge, because it has a great deal to do with attraction. If there are railway lines running nowhere in particular, very few people travel on them, and after a while . . .

RALPH: They have no stations.

RUPERT: Railway companies close them down because they say there's no demand. In this example, and even in your own, Ralph, there is an implication of attractors really attracting.

RALPH: It's a good analogy with Newton's gravitational attraction. With his theory, and equally with general relativity, there are some unresolved difficult problems about action at a distance in space. I think we have the same thing here, but it's a question of action at a distance in time. Two different kinds of time have been confused here.

The train goes down the track and arrives at the station, the attractor, but the problem with thinking of the station pulling the train is that the cause is then in the future. Your argument that the station pulls the train because the people want to go there applies to a different kind of time— time on the longer scale of the evolution of the whole train system. People used to get on and off trains where there were no stations; they just asked the conductors to stop the trains. After many people asked for a certain stop, the railway built a train station there.

The interesting model here is the dynamical model for the evolution of the train system itself, with its various tracks and stations and even with the location of towns and so on. All of this evolves slowly in the course of another kind of time, measured in centuries. This slow evolutionary train also goes toward an attractor, which includes the location of cities, the network of train tracks and stations, and so on. Is this slow train being pulled by the attractor? I think not, because people are exerting their will by getting on and off the fast train wherever they want, and that's the motor of the slow train. The determinant of evolution is free will in the moment, the collective action of citizens in the present.

RUPERT: We face this problem in human psychology. You see, motivations in the ordinary psychological sense are not pushing from behind but pulling from ahead. In courts of law seeking to establish the cause of what happened in a crime, motivation is very important. Did so-and-so willfully murder so-and-so? What was their motive? There's a sense in which a future state or an imagined future state is pulling them. We all have desires and goals that motivate us; we have purposes and aims. All of us at this workshop had the intention of getting to Esalen this weekend, and the intention preceded our coming here. The goal of being here drew our behavior toward it. That goal was in the future.

The concept of morphic attractors in morphic resonance theory, like the concept of entelechy in Aristotle's notion of the soul, tries to deal with the fact that somehow the system, the person, the developing animal, the developing plant, or whatever is subject in the present to the influence of a potential future state that hasn't yet come into being. That potential future state is what directs and guides and attracts the development of the system in the present.

Is that future state existing in the present in some other dimension or direction or time, or is it actually out there in the future pulling from tomorrow or the day after tomorrow through time? There are different ways to imagine how it works.

RALPH: We've arrived again at the imagination.

RUPERT: Exactly. Over to Terence.

TERENCE: I have a lot to say about this discussion, and I'll work backward through it.

Alfred North Whitehead had a phrase, "appetition for completion," which I take to be what this attracting notion is seeking—completion, that is. If we didn't use the word *attractor* and tried to be true to the notion that the process was being pushed from behind, we would have to use a word like *propeller* or *motivator*. These seem to me, intuitively, to be inelegant terms. They immediately raise questions of operational detail that *attractor* doesn't. We know when things are attracted to something; they simply move toward it. If something is "propelled" toward something or "motivated" toward something, we have to visualize it strapped to an engine that is moving it toward an end state, which, somehow it is able to magically locate.

If you view the attractor as the bottom of an energy well, then anything put into the energy well will make its way to the attractor because the attractor is the least energetic state. The whole system naturally tends to move in that direction. The idea that the cause is in the future makes hash of the conventional notion of causality, so it's something that science is very keen to discredit. The backwash from the acceptance of this idea would make the practice of science much more difficult.

For many years, Ralph and his colleagues have been modeling plant growth, dripping faucets, coupled oscillators like groups of cuckoo clocks hung on the wall, and this sort of thing. The modeling challenge for the future is human history. We will no longer be playing little games to demonstrate something to a group of students or colleagues, but we will actually be proposing models and methods powerful enough to begin to model the real world. These models will deal not only with the real world of biology, but with the real world of the felt experience of being embedded in human institutions.

I think the whole reason history has been bogged down during the twentieth century is because of an absence of belief in an attractor. The legacy of existentialism and the philosophies constellated around it is the belief that there is no attractor, no appetition for completion. Everything is referent to the past up through the present and goes no further.

My tendency is to carry any principle to its ultimate extrapolation. In thinking about complexity in relation to falling temperature I glimpsed something I had previously overlooked.

If, in fact, the increase in complexity in life is directly related to falling temperatures in the universe, then it seems reasonable to suppose that the most complex states in future cosmic history will occur at very low temperatures. It's interesting that a phenomenon like superconductivity, which is fascinating to solid state engineers as a way to preserve information from decay, occurs at low temperatures. If you put information into a superconducting circuit operating at around absolute zero, it's impossible to disrupt that circuit without destroying it. As early as the mid thirties, people like Erwin Schroedinger suggested that, since life seeks to stabilize itself against mutation, the obvious principle to aid in that task would be something very much like superconductivity.

In fact, the way in which charge transfer and things like that occur in DNA suggests that nature may have incorporated this principle into its mechanics. What this tells us in the present is that our current cultural phase transition vis-a-vis machines may signify that we are not, as I've always thought, very close to the maximized state of novelty. Rather we may be somewhere out in the middle of the topological manifold I call the "novelty wave," which goes from the beginning to the end of all things. The cultural transition that we are experiencing is a downloading of all novelty so far achieved into a much colder and stabler regime of silicon crystals and arsenic-doped chips and this sort of thing. This is a fairly appalling idea, because we all have a horror of being replaced by machines. On the other hand, procaryotes were replaced by eucaryotes, and there have been several other replacement scenarios in the history of life.

This point about cooling and complexity seems to imply, in my own theory of the "time wave," that the zero point attractor at the end of time may in fact be the absolute zero point, and that what the time wave or the fractal time manifold really describes is the fluctuation of the career of heat throughout the life of the universe. In domains of high heat, information is degraded and novelty is lost, and there is a kind of recidivist tendency. When temperatures fall, order reasserts itself and things stabilize.

In the wake of each ice age, human populations emerged with better tools, better languages, and better techniques than before. It was as if the increased environmental pressure, and perhaps even the increased need to spend more time together, synergized the emergence of higher states of order.

We associate lower temperatures with death. We all understand that if temperatures drop below a certain very narrow range, that's it for us. The machines we are creating, however, are operating more and more efficiently as temperature is dropped. In the realm of absolute zero, almost miraculous things can be imagined in the way of technical storage and retrieval of information.

RALPH: Terence, what is the optimal environment for biological information storage?

TERENCE: A very cold regime is optimal for mushroom spores. The actual expression of the spores' genomes through the growing of fungi has to occur in a normal biological regime, but spores stored in liquid nitrogen can be stored indefinitely. In fact, most tissue can be stored indefinitely at these low temperatures. It's not very interesting to be at 70° Kelvin, but it is the path toward a kind of immortality because that's where preservation takes place.

RUPERT: The idea of information storage at low temperatures is interesting when we consider the difference between spoken and written language. The first written languages we know about were written on rocks, the ultimate, low-temperature, crystalline storage system. For example, the Ten Commandments given to Moses were on tablets of stone. Writing on rock is a kind of permanent storage system. Putting things in silicon crystals is more sophisticated but is still essentially a low-temperature storage method. You can't write on water or on the wind.

Written language creates the illusion for us of an independent world. The notion of a transcendent eternal world of Forms couldn't have arisen until written language did, because written language provides the model for it. By what I think of as a kind of idolatry, human-made symbols and structures, when written down, are imagined to endure forever in some other realm. Spoken language is far older than written language, but is a process that happens in time. The memory involved in oral cultures is carried in stories that are continually retold and that evolve as they are transmitted. The spoken record, the story, develops organically as time goes on, and there's nobody around to say, "Well, you've got the story wrong; in the book it's written like this."

Spoken and written language provide us with different models of reality. Oral tradition has an evolving and yet conservative quality, and suggests a model of reality rooted in habit and tradition yet open to the creative imagination. The model of written language projects the idea of things being fixed by being written down and gives the impression of a realm of eternal Forms or formulae.

RALPH: I imagine, just to be contrary, that mathematics probably preceded not only writing but language as well. Certainly mathematics preceded writing. In mathematics there are, for example, the circle and the line, which were, for Plato, ideal, eternal Forms. Do we need writing on stone to think of a line or a circle or a triangle as being an eternal Form? The evolution of this kind of mathematics was probably achieved by people drawing in sand. Writing evolved from this drawing in sand, and only later did we begin drawing on stone. It's possible that the idea of eternal Forms, laws, and so on emerged before writing on stone, and that writing on stone was just a concretization of those ideas. This suggests a migration in evolution from the immaterial to the material, from the abstract to the concrete, which is opposite to what a lot of people think.

TERENCE: To summarize this dialogue, the concept of the attractor was introduced and explored in detail, probably to the clarification of at least two of the participants, Rupert and myself. The role of the attractor seems central to understanding what chaos dynamics is offering that is new. Rupert dwelt on the very useful analogy of the way in which order is attained through phase transition. He chose the model of cooling, both in the specific case of the cooling of a liquid and in the case of the whole cosmos as a slowly cooling solution going through phase transitions from lower to higher states of order as the temperature falls.

Ralph then raised the stakes in an attempt to communicate the complexity of what chaos dynamics is saying. He went beyond the notion of the attractor and the basin of attraction and introduced the idea of bifurcation, which is a further development in the metaphor. The fruitfulness of, aesthetics of, and constraints on the process of modeling were discussed, as was the way in which modeling allows us to build up provisional pictures of reality into which we can establish correctional feedback loops. These feedback loops allow us to navigate toward ever clearer images of the system we have targeted for modeling. Also touched upon were ways in which new forms emerge out of the natural order and ways in which they are stabilized in time.

The longer we talk, the more creation, imagination, and chaos all seem to be the same thing, and there's a kind of melding of these concepts. One can play any role and find that it's very much like the role one has just left behind. This means we are succeeding; that the separate notions we each represent are annealing.

Repression of chaos results in an inhibition of creativity and thus a resistance to imagination. The creative imagination, manifested most profoundly by people like Euler or Bach, should be functioning in everyone. People have a resistance to their own creative imagination.
    —Ralph Abraham

The imagination argues for a divine spark in human beings. It is absolutely confounding if you try to see imagination as a necessary quantity in biology. It is an emanation from above—literally a descent of the world soul into all of us.
    —Terence McKenna

One thing that's clear is that chaos is feminine, and creation out of chaos is like creation out of the womb, an all-containing potentiality emerging out of darkness.
    —Rupert Sheldrake

3

CHAOS AND THE IMAGINATION

RALPH: There's a resonant relationship between chaos on plane three and chaos on plane one of figure 3. My appreciation of this resonance has increased over the past years. I am a devotee of chaos, but not only on the mathematical plane. I should like now to speak about chaos in ordinary life and the relationship of this chaos to the imagination. Along the way, I'll try to explain why I think that chaos theory is the biggest thing since the wheel. Terence has emphasized the importance of chaos and fractals in his own thinking. I'll provide a further reason to regard the chaos revolution as good news.

Somebody came to me last year and gave me a splendid quote, knowing I was writing a nontechnical book on Chaos, Gaia, and Eros: "All creation begins in chaos, progresses in chaos, and ends in chaos." The word *chaos* occurred for the first time in Hesiod, around 800 B.C., at the beginning of the Orphic tradition of ancient Greece. The word appeared in his *Theogony*, which was about the creation of the gods and goddesses one by one. The three main deities were Chaos, Gaia, and Eros.

This first time the word appeared in literature, it had nothing to do with what we now mean by chaos in the English language and in ordinary life. At that time, it meant a sort of gaping void between heaven and Earth out of which form emerged. Creation came out of chaos, but chaos did not mean disorder or anything negative; it only meant a gaping void.

Proper nouns such as these in Greek literature don't mean people or humanoid gods; they're abstract principles, particularly in Homer and Hesiod and in the Orphic literature. The three I am discussing—Chaos, Gaia, and Eros—are cosmic forces: Chaos is a sky concept; Gaia is an Earth concept; Eros is a spirit concept. Whereas Chaos and Gaia are feminine in Hesiod, Eros is bisexual, an androgyne. These are abstract concepts of the sky, the Earth, and the creative tension in between them;

they are like the body, the soul, and the spirit of the world.

The most obvious feature in the sky is the Milky Way. When it is overhead, it appears as a kind of gap between the two sides of the sky. The sky is sort of like an Easter basket with the ecliptic around the waist, like a snake. All that is below the waist is considered the Underworld. All above it is the Overworld, or ordinary reality. There is a handle across the top that is the Milky Way, which passes right through the middle of the Overworld as the royal road for the gods traveling between the Underworld and the Overworld.

The Milky Way looks like a fractal photograph, the prototypical chaos in the sense of randomness or disorder. Understanding this, we can connect the word with an earlier word, a proper noun, and cosmic concept that represents what we see in the Milky Way itself. That noun is "Tiamat," the goddess of chaos from *Enuma Elish*, the great epic poem of Babylonian literature. This poem is a creation story about the origin of the gods and the creation of the world by Tiamat and Apsu, the goddess and god of chaos who lived in the water. These two deities created the pantheon of gods and goddesses and the whole world. Later, Marduk, a younger god in the next generation, appeared, and there was conflict between the older and the younger gods over how the world ought to be run. Apparently this conflict represented a social transformation, the demotion of Apsu and Tiamat from the pantheon of the city of Babylon. Marduk became the main god, Mister Big, of Babylon, around 2000 B.C., coinciding with the sweep of patriarchy over that city, propelled on a new type of war chariot with spoked instead of solid wheels.

TERENCE: You've reached the wheel.

RALPH: The wheel itself, as a mathematical model, is the paradigm of order. Order has come to mean a process that is either static or periodic, regularly changing in a cycle. In short, according to the Babylonian epic *Enuma Elish*, Tiamat (chaos) was killed, ripped to pieces to create a new world order by the hero of Babylon, Marduk. His New Year's celebrations were honored all over old Europe, including even Stonehenge. At the New Year's festivals, the epic poem was read as an annual reminder that chaos was bad and was killed and replaced by an order associated with periodicity, cycles, the wheel, perfect roundness, and

so on. For two thousand years, this poem was read every year in a celebration lasting eleven days called the Akitu festival in Babylon. There
were almost identical festivals in Egypt, Crete, and Canaan.

All of this reveals the importance of today's chaos revolution.
Chaos is recovering from being banished to the unconscious since around
2000 B.C. or so. It has suffered four thousand years of repression. To this
day, in our culture, we think we have to watch out for chaos; it has to
be replaced by order. Scientists most especially hate it. Today, Tiamat
finally has to be accepted as a friend and reinstated upon her throne.
This is big news.

Now, about the imagination. So far in our discussions, we have
developed the connection between creativity and the imagination and the
one between chaos and creativity; thus, we have made an indirect connection between chaos and the imagination. I'm going to try to make a direct
link between chaos and the imagination, as shown in figure 4.

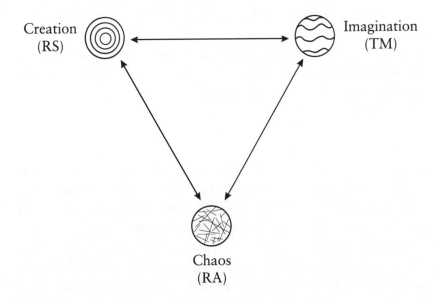

Figure 4. Our Trinity. *The triangle of roles adopted by the authors for the first
four trialogues.*

Repression of chaos results in an inhibition of creativity and thus a resistance to imagination. The creative imagination, manifest most profoundly by people like Euler or Bach, should be functioning in everyone. People have a resistance to their own creative imagination. I'm suggesting that this resistance has a mythological, ritual base. It represents a very deep chreode, a runnel in the morphogenetic field with a depth of three to five Waddingtons.

TERENCE: Waddingtons?

RALPH: Rupert and I wanted to quantify the depth of a runnel. Since C.H. Waddington introduced the chreode concept in connection with embryology, as Rupert has explained, we thought it appropriate to take his name for a unit of depth in the morphogenetic field.

Habits, according to Rupert, are runnels. They are deeply worn paths in the morphogenetic field. Habit itself is a habit. We have a habit of having habits. The habit of habits has a mathematical model called by Waddington a chreode. This chreode is an attractor and its basin (figure 1). Any state in the basin moves toward the attractor. The attractor is the teleology of its habit. The habit of habits is the habit of living in a landscape with peaks and valleys where there is a tendency to go down the valley, down the track to the station. The habit of habits is good, but some habits might be harmful.

One thing that creates a really deep runnel is the repetition of a myth for eleven days in a row, once a year, for two thousand years. That makes the myth a more serious matter, a deeper rut. As an example of a really deep chreode, therefore, I recommend for your consideration the New Year's festival of Akitu. Old Testament scholarship has produced a comparative study of New Year's festivals around the world that provides amazing evidence of morphic resonance on a planetary scale. Based on this evidence, there was, before radio, a global resonance with the Gaian mind on the same level as the distribution of the so-called Venus figurines of the Gravettian culture all over the planet.

This is the link between chaos and the imagination. The repression of chaos, and the resulting obstruction of imagination, has led to the development over thousands of years of the serious, perhaps fatal, ecological problems that we have today. The growth of this problem beyond all bounds, with few people even trying to do anything about it, suggests denial and a lack of imagination. This is an artifact of a cultural flaw that came along with the patriarchy.

I'm not saying that patriarchal culture is bad because it's patriarchal. I'm saying it's bad because it has repressed chaos. Patriarchy has made chaos bad and it has made Marduk boss: the god of law and order. We must reject this view of chaos so that the planet and life and love can be saved. Now, lo and behold, an event has come along that is as positive, perhaps, as the discovery of the wheel was negative. Without giving up the wheel, or our automobiles, we are regaining chaos for potential partnership with the wheel. Chaos and order. Chaos and cosmos. Chaos and the imagination.

TERENCE: Chaos and the imagination are paradoxically co-present in everyday life, in the dimension in which we find ourselves. I would like to expand on what Ralph has been saying from my own perspective. Chaos is feminine. Chaos is intuitional. Chaos has a very flirtatious relationship with language. The process of creating a culture has to do with how we relate to the seduction of chaos. Since we're talking about everyday life, it seems reasonable to talk about how we come to be in the kind of planetary mess that we're in and whether these extremely rarefied and abstract matters that we've been talking about can in any way be brought tangential to the issues of a burning planet.

The first key is the power of chaos, which is feminine, beyond prediction, and beyond full, rational apprehension. The second source of power is the Divine Imagination, the imagination that is our richest legacy, the birthright that connects us to the divine. It's our poetic capacity, our ability to resonate with a notion of ideal beauty and to create that which transcends our own understanding in the form of art.

What does all this say about the situation we're in and how we've gotten here? I believe that the importance of psychedelics relates primarily to this situation, and not simply because they synergize cognition.

We have a secret history. Knowledge has been lost to us and is only now recoverable in the light of the mindset possible if we accept the new chaos paradigm. How this secret history relates to the Gaian mind and the world soul is a kind of new revelation. We must become aware of the astonishing fact that as a species we are the victims of an instance of traumatic abuse in childhood. As human beings, we once had a symbiotic relationship with the world-girdling intelligence of the planet that was mediated through shamanic plant use. This relationship was disrupted and eventually lost by the progressive climatic drying of the Eurasian and African land masses. This was the literal force behind our fall into history, our expulsion from Eden. All the primary myths of the Golden Age found and lost point to the fact that once we lived in dynamic balance with nature—not as animals, but as human beings in a unique way that we have lost. How have we lost it, and what have we lost?

Psychoactive compounds that were brought into the expanding human diet during the early part of our evolution inhibited the formation of the ego, promoting instead collectivist, tribal partnership values operating intuitively in a reciprocal feedback relationship with the feminine vegetable matrix of the biosphere. In other words, nothing was verbalized and everything was felt. Everything was intuited. Regularly, at the new and full moons, small groups of hunter-gatherers and pastoralists took hallucinogenic plants and dissolved their boundaries, engaging in group sex and annealing irregularities that had cropped up in their personal self-imaging since the last session. These practices kept cultural reality grounded on the plane of group and species values, which was in dynamic balance with the ecosystem.

When these practices were disrupted as supplies of these plants diminished, new religious forms arose, and the time between these great festivals grew longer and longer. The ego began to take hold, first as a kind of cancerous aberration, then quickly becoming a new style of behavior that eliminated other styles of behavior by suppressing access to the sources of chaos.

The point I want to make is that between the ego and the full understanding of reality is a barrier: the fear of the ego to surrender to the fact of chaos. In a premodern society, no woman could escape chaos because of the automatic birth script that had women giving birth over and over and over until death. Women are biologically scripted into being much closer to chaos simply because there are certain episodes in the life of a female that are guaranteed to be boundary dissolving. The psychology of feminine sexuality, which involves the acceptance of penetration, creates an entirely different relationship to boundary than does the male need to fulfill the potential to penetrate.

We have lost touch with chaos because it is feared by the dominant archetype of our world, the ego. The ego's existence is defined in terms of control. The endless modeling process that the ego carries out is an effort to fight the absence of closure. The ego wants closure. It wants a complete explanation.

The beginning of wisdom, I believe, is our ability to accept an inherent messiness in our explanation of what's going on. Nowhere is it written that human minds should be able to give a full accounting of creation in all dimensions and on all levels. Ludwig Wittgenstein had the idea that philosophy should be what he called "true enough." I think that's a great idea. True enough is as true as it can be gotten. The imagination is chaos. New forms are fetched out of it. The creative act is to let down the net of human imagination into the ocean of chaos on which we are suspended and then to attempt to bring out of it ideas.

My model for the psychedelic experience is the night sea journey—yourself as the lone fisherman on a tropical sea with your nets. Sometimes, something tears through your nets and leaves them in shreds, so you just row for shore and put your head under your bed and pray. Other times, what slips through the nets are the minutiae, the minnows of this icthyological metaphor of idea chasing. Sometimes you actually bring home something that is food for the human community, from which we can sustain ourselves and go forward.

We haven't talked much about art and aesthetics so far, but I think a unique characteristic of the human world is its appetition for beauty. This is another place where Platonism shines radiantly. Plato held that the Good was the Truth and that both were the Beautiful. This is a very quaint idea from the point of view of modern philosophy, but I feel it in my bones when I actually connect myself up to the planet.

I believe chaos is capable of being a tremendous repository of ordered beauty. There is no disorder in the old definition of chaos. Chaos as disorder is a kind of hell notion, a kind of hypostatization of an ultimate state of disorder. Nowhere in the world that is deployed through space and time do we encounter this. Instead, we encounter embedded order upon embedded, fractal order.

Finally, for me, imagination is the goal of history. I see culture as
an effort to literally realize our collective dreams. The process is now
operating on a very crude level: You make your mask and I make my
mask and then we dance around together. You design your shopping mall
and I'll design my World Trade Center and we'll put them on the same
piece of real estate. Today, through media, psychopharmacology, virtual
reality and human/machine integration, we're creating a situation in
which the imagination is something that we can share. The path of mind
through its own meanderings will become something that can be recorded and played back. We will have the possibility of living in our
own past, or of creating and trading realities as art. Art as life lived in the
imagination is the great archetype that rears itself up at the end of history.

The imagination is an auric field that surrounds the transcendental
object at the end of time. As we close distance with it, all of our cultural
expression and self-awareness takes on a curiously designed quality.
The world is very heavily designed in a way that it never was before.
Morphogenetic fields of great size and scope, in the form of international
schools of architecture and design, touch whole continents. Entire cities
are given certain ambiences. This is the summoning of imagination into
the human scale. It's like a god that we call down and draw to Earth.
William Blake called it the Divine Imagination. It's the four-gated city,
the flying saucer. We are on a journey to meet the great attractor, and as
we close the distance it is more and more a multifaceted mirror of our
own images of beauty. This journey is an ascending learning curve that is
becoming asymptotic; at that point, we are face to face with a living
mystery that is within each and all of us.

The imagination argues for a divine spark in human beings. It is
absolutely confounding if you try to see imagination as a necessary
quantity in biology. It is an emanation from above—literally a descent of
the world soul into all of us. We are the atoms of the world soul. We open
our channel to it by closing our eyes and obliterating our immediate, per-
sonalized, space-time locus. We then fall into the imagination, which
runs like an endless river through all of us and is driven by the hydraulic
momentum of the cataracts of chaos.

Riverine metaphors are endlessly applicable. They represent the
flowing of forces over landscapes, the pressure of chaos on the imagina-

tion to create creativity. These things are icons for the world that wants to be. The key is surrender and dissolution of boundaries, dissolution of the ego, and trust in the love of the Goddess that transcends rational understanding. There will come a moment that will be an absolute leap into faith, and we will simply have to believe that something is waiting there because the dominator style of the ego has left us no choice.

RALPH: "Cataracts of chaos"—these words roll off his tongue, which obviously has a direct connection to the cataracts of chaos themselves! There are so many different ideas to respond to here. I'll try to limit myself to one or two.

I'd like to connect two statements with a key from the Eleusinian mysteries. This conversation has stimulated speculation on what happened to the cataracts of chaos in the hands of the patriarchy.

Historians generally suppose that there's a straight line of influence from Babylon to Ugarit to Minoan Crete to Mycenae to Athens, a cultural diffusion that includes the festivals, which, as I said, are very similar. However, in the Babylonian festival, by 2000 B.C., there was already extensive domination by the patriarchal male god of order. In Minoan Crete, on the other hand, according to the excavators and all who have examined the artworks that remain, there was an outstanding, longlasting florescence of partnership culture with no domination by a male god. This presents a little question mark about the diffusion of culture from Babylon to Crete.

There is a complementary theory that postulates there was a renewal of partnership culture in Crete that came from India in the early days of Cretan culture, around 3000 B.C. In any case, it is agreed that there was diffusion from Crete to Greece and that the last remainder of Cretan culture, the last vestige of the Garden of Eden we're talking about, was a tremendous happiness and florescence of beauty in all aspects of life.

The Eleusinian mysteries were said to be identical to those that in Crete were celebrated publicly. They only became secret after the patriarchal takeover by Mycenaeans in Crete and the exportation of this dual culture to Greece. There occurred a bifurcation into the Overworld and the Underworld, the conscious and the unconscious. The culture of the Goddess, and the partnership of gods and goddesses in Crete, continued to exist in Greece, but as secret festivals known as the mysteries.

Among the common aspects of the festivals in Crete and Eleusis were the sexual and psychedelic rites celebrated in rigidly specified ritual formats on an annual basis. In patriarchy, because of its patrilineal aspect, one has to know the father of the children, not just the mother. Therefore, monogamy is very important to patriarchy, and this was in strong conflict with the Cretan rituals. Sooner or later, the sexual rituals had to go.

This is a simple explanation, not confirmed by scholarship, for the repression of these rites. The Eleusinian mysteries disappeared, and what we are left with is the paradigm of Marduk the control freak, law and order, the repression of chaos, and the constipation of the imagination and fantasy.

Somehow, sexual license, the psychedelic ritual, and chaos go together. The New Year's festival continues to our day, but the sexual/ psychedelic/chaotic aspects of it were long ago removed and replaced with a recital of the conquest of chaos by the god of order.

TERENCE: Do you think it's because the ego is basically on a control trip and must become the center of the space in which it operates? Egomaniacs control women and resources. They believe vigilance is the key, as in the national motto of Albania, "Fear it is that guards the vineyard."

Our culture definitely takes an egocentric dominator view. The fear of the psychedelic experience is quite literally the fear of losing control. Dominator types today don't understand that it's not important to maintain control if you are not in control in the first place.

In a tribal society, for instance, where there is no property and there are no assigned sleeping places, everyone behaves according to something that we can only call whim. Whim has been replaced by "mine, yours, my spot, your spot, my food, your food," which has led to the dissolution of the psychedelic collectivity. The world has become very threatening and everything is up for grabs. Now we have to rely on the interpretation of group values instead of on felt intuition. The ego is paranoia institutionalized.

RALPH: I guess it's a package deal. We have the dominator, the patriarchy, and the ego, which in our culture we can see as a male disease like testosterone poisoning.

TERENCE: I don't see it as a male disease. I think everybody in this room has a far stronger ego than they need. The great thing that Riane Eisler, in her book *The Chalice and The Blade*, did for this discussion was to de-genderize the terminology. Instead of talking about patriarchy and all this, what we should be talking about is dominator versus partnership society.

RALPH: I agree with you; it's a good service she has done. However, the problem we're faced with is how to get back to the partnership paradigm.

One practical suggestion is your missionary appeal for psychedelic usage. The ego aspect of the problem arose not only through the suppression of psychedelic usage in ritual but through the gradually increasing interval between festivals. This process was similar to the cooling of the universe, out of which different material forms crystallized. The dominator society sort of crystallized out through the gradual increase in the interval between partnership rituals.

TERENCE: That is part of it, definitely. The phonetic alphabet plays a role of further abstracting from process, giving permission for all kinds of curious dis-ensouling maneuvers on the part of the dominator/ego.

RALPH: Well, we still have mathematics and music.

TERENCE: Yes, but so few people have them. And those few who do are the creative engines of their societies. They keep the connection to the muse.

RALPH: Do you think that the Eleusinian mysteries could be reinstituted? Or has this possibility been irrevocably lost to "progress"?

TERENCE: The psychedelic revival is an effort to find our way back to something like the mysteries. We are not the first nor the most eminent to suggest this kind of reengineering of the human animal. I call attention to the words of Arthur Koestler, the great anticommunist freedom fighter and scientific intellectual. In a book called *The Ghost in the Machine*, he concluded that there has to be mass pharmacological intervention to change human behavior. He envisioned a drug that inhibits territoriality.

Our reflexes and our mental set are highly and well adapted to the stoning to death of woolly mastodons, but, since we so rarely do that, we need to retool for living in peace while managing limited resources. The dissolving of boundaries by psychedelics certainly makes them candidates for antiterritoriality drugs.

RALPH: Do you think that some of our existing national holidays could be changed, or that a mythological mutation could be introduced that would go in this direction? For example, in Switzerland, they recently invented Fastnacht. They previously had no rituals at all, barely Christmas, and there was a suffering of enormous boredom among people there. It was said that there was no known way to make a new friend in Switzerland. So, just a few years ago, they instituted Fastnacht, in February. It is three days and nights of alcoholic revelry around the fantastic reenactment of a medieval drama. It involves people marching in the streets in parades led by musicians who have practiced a medieval song on medieval instruments all year long just for this three-day ceremony. Now it is said that, during Fastnacht, you can make a new friend.

TERENCE: Something along that line that I've advocated—sometimes facetiously, sometimes seriously—is calendrical reform, and I have just the calendar all worked out. I won't lay it all out here, but the basic notion is that it's a lunar calendar of thirteen lunar cycles. It has three hundred and eighty-four days, and consequently it precesses nineteen days against the solar year. This would have the effect of taking the great yearly events of the calendar and slowly moving them through the seasons. For instance, if we kept Christmas on December 25, and you as child celebrated Christmas in winter, then as a teenager you would celebrate it in spring, and as a young adult you would celebrate it in high summer. As an older person, it would occur in autumn, and then, when you were truly old, Christmas would return again to the winter.

The notion is to overcome the really bad dominator idea that the calendar should be anchored rigidly at the equinoctial and solstitial points so that the heliacal rising of the equinoctial sun is always in the same place. Our current calendar sends the message that there is stability. The calendar is the largest framework there is; in it, all other contexts are somehow subsets. The solar calendar is an effort to deny humanity's mortality by reinforcing a false notion of permanence. What we actually want is a calendar that says to us, "All is flow; all is flux; all relationships are in motion to everything else." This is a truer picture of the world. This may seem trivial, and exemplary of why we eggheads are harmless. But, think about it.

If we yield the structure of the calendar to the dominator culture, letting it tell us what kind of calendar we shall have, then we shall all live within the context of the dominator framework. Changing the calendar would have tremendous consequences, and it would not be opposed by the dominant culture until it was too late. It would be regarded as some kind of a crank thing, because it wouldn't be realized that we were twiddling with the dials of our whole civilization's image of time and change.

The year 2000 provides a built-in opportunity to switch the train to a new track because at these millennial moments there's a certain uncertainty in the mass mind about how to proceed. If you just jump up onto the stage and say, "This is it, folks!" you might be able to pull it off.

RUPERT: The first thing that occurs to me is that your idea of a lunar calendar in which the months and festivals move around the year is already in place. It's the calendar of the Islamic world. For example, Ramadan, the fasting month, retrogresses around the year, so people experience it both in winter and in summer in the course of their lives. This is not exactly a confirmation of your theory about such a calendar breaking the dominator mode.

TERENCE: Your point is certainly, um . . . unwelcome.

RUPERT: I'm still baffled by chaos. Its meaning keeps shifting. We started off with chaos not having a negative connotation and just meaning the yawning void. In fact, as Ralph explained, it was the sky. Then it was the great womb, the source of all things. It then got turned into Tiamat, who was slain by Marduk. In the first chapter of the book of Genesis, it's the "deep." The "spirit moving on the face of the waters" sounds to me like a wave theory of creation, as wind on water sets up waves. One thing that's clear is that chaos is feminine, and creation out of chaos is like creation out of the womb, an all-containing potentiality emerging out of darkness.

One of the metaphors that Terence used of the imagination was the dipping of nets into the deep ocean and the pulling up of coelacanths of the imagination. This corresponds to the kind of imagery used by the Jungians for the unconscious bubbling up from below, and it also fits with the idea of creativity welling up from the darkness of the Earth. Again, this is a model used in the book of Genesis. The biblical account of Creation doesn't have God creating animals and plants. It has God saying, "Let the Earth bring forth fresh growth; let there be on the Earth plants bearing seeds, fruit trees bearing fruit, each with seed according to its kind" (Genesis 1:11), and then, "Let the Earth bring forth living creatures according to their kind" (Genesis 1:24). The Earth brings them forth from herself.

Terence also used the metaphor of the imagination descending from above, which is a traditional Platonic or Neoplatonic image. In this model, creativity comes down from above, becoming more and more manifest through a series of stages. These metaphors have given us both a top-down and a bottom-up model of creativity. Most theories of creativity I come across oscillate unstably between those two models. I usually try to resolve the conflict by saying it must be a mixture of both.

In the realm of theology, these two models are known as ascending and descending Christology. In the first three gospels, the model is Jesus Christ, who was born as a child, initiated by John the Baptist, and received a new spiritual illumination at his baptism. He underwent a development and spiritual transformation and became God. A man becoming God is an ascending or bottom-up developmental process. Then, in the Gospel of John, there is the Platonic model of the Word becoming flesh, God becoming man, the top-down model of creativity. These models have always co-existed in Christian theology in a dialectical tension.

In most discussions of creativity, one gets into the same polarity. Chaos is seen by Ralph on the one hand as a chaotic, indeterminate process that, in some sense, liberates us from older models of control and mechanistic determinism. On the other hand, he also adopts the top-down model because he wants the generation of chaos to come from simple mathematical principles. There is also the attempt to tame chaos by modeling using the top-down method. Mathematical modeling of chaos, if not exactly in the dominator mode, is still, I think, within the Saint George and the Dragon archetype. Actually, Saint George didn't slay the dragon. He pierced the dragon, tamed it, and led it captive into the city. The dragon in that myth is obviously another form of the primal monster of chaos.

Are these metaphors just alternative models between which one can switch, like changing tracks? Are they aspects of the same process, or are they complementary processes?

TERENCE: I think they're different pictures of the same process. Much as the image of the Star of David can be seen as the interpenetration of two triangles, this process is the notion of "as above, so below." We don't understand it unless we're somehow able to hold both images simultaneously. It's a cardinal premise, I think, that in talking about these kinds of things you can't force closure. This is alchemical thinking, in that the things that are being described are multidimensional objects that can sustain seemingly contradictory descriptions. These are compound, complex concepts that must have this overlay to be correctly communicated or appreciated.

RALPH: As above, so below. In the history of the chaos concept, as I described it, there is a syncretism of the Hesiodic Chaos concept as the Milky Way, a yawning, celestial void, with the Babylonian Tiamat concept of chaos as a dragon or sea serpent. There are numerous pictures in Babylonian art of this sea serpent with a bridle and Marduk standing on its back holding the reins, driving it along. Marduk has conquered the sea serpent. For some reason, mythology at this point in history required that a celestial figure be overlaid on an Earth figure. There are many gods and goddesses in the pantheon, and frequently they are different aspects of the same thing. Rupert took me into a church in Santa Cruz and showed me a shrine to Our Lady, the Virgin of Guadalupe, with a painting of her as a black goddess, a chthonic Earth Mother figure. She's wearing a dress with stars printed all over it.

In the Easter basket model, the sky as a visible hemisphere comes to an end on the horizon where the Earth begins. The connection between the sky and the Earth is actually the Milky Way, the handle of the basket. This is the royal road of the gods down which Orpheus goes to the Underworld to look for Eurydice. The point where the sky meets the Earth, where the handle of the basket connects to the woven part, is where the mathematical model meets the unconscious of the Gaian mind. This is where the mathematical version of chaos meets the chaos of everyday life, and where the erotic and synergistic relationship between the Earth and sky versions of chaos is going to take place.

TERENCE: Eleusis was a great turning point and a cultural episode not frequently enough discussed. After thousands and thousands of years, the Goddess-worshiping, orgiastic, psychedelic religion finally was confined to a few shrines in Greece and Crete. Then, ultimately, it was confined to a few shrines only in Greece. It lingered there until Alaric the Visigoth finally did it under.

This boundary-dissolving relationship to the vegetable, Gaian mind left our tradition only about seventeen hundred years ago. In that seventeen hundred years, in the absence of a dialogue with the Gaian expression of chaos, successively more deadly cultural forms, beginning with the phonetic alphabet and moving on to movable type and all the rest, evolved. Each one of these technologies has had a tremendous negative impact on our self-image and has entangled us deeper and deeper in a kind of Faustian pact with the physical world. It's that blindness that has led us to the present situation. In the absence of any boundary-dissolving ecstasies, we are left with the machinations of the ego, which has led very quickly into a cultural cul-de-sac from which there may or may not be an escape.

In studying that wrong turning, what was betrayed, what came out on top, and what was suppressed, we can perhaps run the film backwards and in some sense restore the previous situation. This involves opening our lives to chaos and becoming much more a part of the will of the world soul. It means recapturing the Greek sense of fate that has been replaced in our minds by the Faustian illusion of control and dominance.

RALPH: The good news is that this opening to chaos is actually under way. The chaos revolution now taking place throughout the sciences is a major setback for the forces of law and order, control and dominance. Scientists, the high priests of Marduk, must now accept chaos and replace Tiamat on her rightful throne. This is why I say that chaos is the biggest thing since the wheel. Imagination, creativity, and inspiration are all on the upswing. Chaos, Gaia, Eros—arise!

I'm into the world soul as the largest and smartest creature imaginable. It isn't the God who hung stars like lamps in heaven, and it isn't the force that spins the galaxies on their axes. The world soul is something that has arisen out of biology. It's an organism within the universe of space and time, but it didn't make the universe. It's an inhabitant of it, but on a scale that makes us mere atoms within its form.
   —Terence McKenna

The Gaian mind may be faltering due to a bad habit. Incarnation is addictive, and the reason there is an infection out of control on this planet is because of this bad habit.
   —Ralph Abraham

Insofar as the mushroom and the human psyche have had a symbiotic relationship, the mushroom-induced experiences in human consciousness are in the morphic field of this symbiotic relationship. Therefore, these experiences could be carried by the mushroom.
   —Rupert Sheldrake

4

THE WORLD SOUL
AND THE MUSHROOM

RUPERT: Terence, do you believe that creation comes out of randomness?

TERENCE: It seems as though randomness is the least likely thing. The probabilistic view assumes that nature has randomness built into it because randomness is something that probability mechanics is competent to make statements about. Yet nowhere in nature do we encounter this randomness. The averaging of many data samples suppresses the characteristics of unusual events by giving smooth curves that we assume describe processes we've never actually examined with sufficient care. When we look carefully, as we are beginning to, we find fractal structures that are self-similar and surprisingly unpredictable.

My notion of a novelty wave is offered in this new spirit. It would replace the idea of time as a smooth surface—as in the pure duration of the Newtonians—with the idea of time being composed of discrete elements complexed together in a unique way under the aegis of definable rules. This can be visualized as a fractal wave pattern overlaying the universe of space/time. The pattern imparts to time the local quality of either conserving accumulated material from the past or expressing novel connections with emergent properties previously only implicate.

RUPERT: If there's no randomness in the universe, then what do we mean by chaos?

RALPH: Well, ironically, in the chaos revolution, in the technical jargon of mathematics, we're talking about mathematical models that are chaotic in that they share the intuitive meaning of chaos, yet they are not random. Originally, when these models were discovered, they were called strange attractors. Then we saw that they were ever-present in nature, so we decided they shouldn't be called strange. Someone suggested they ought to be called chaotic attractors. There was objection on the grounds

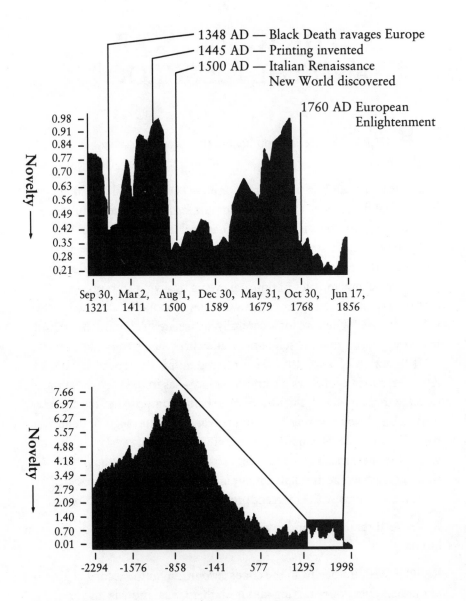

Figure 5. The Time Wave. *A theoretical construct developed by Terence McKenna that attempts to plot the ebb and flow of novelty through time. Novelty increases as the wave meanders toward zero. A period of nearly 4,300 years appears in the bottom portion. The upper portion shows a blowup of nearly five hundred years' duration. As the values approach zero, novelty and density of connectedness increase while entropy decreases.*

that these models were the opposite of what chaos ordinarily means, because they have order. I understand Rupert's question to refer to random behavior that is outside the realm of these mathematical models, which are chaotic, yet not random. The question is whether anything in nature lies outside the "ordered" realm of the chaotic attractors.

RUPERT: Yes.

RALPH: In my view, there is a lot left beyond our ken. We don't know if, in the future, models for this "lot left beyond" will emerge or not. Given sufficient time to continue our present evolutionary path with science, mathematics, computers, and so on, I should think that the amount of so-called "random" behavior—in the sense of our ignorance of any structure in it—would be in sharp decline. Nevertheless, in the world of these mathematical models for chaos, there is space for novelty, mutation, and the discovery of totally new patterns. I don't think that we need randomness in order to have the evolution of new forms.

RUPERT: What we do need is a universe sufficiently open and undetermined for new forms to have space to arise in. If all of nature is already geared up to follow predetermined waves, patterns, forms, and so on, whether they're modeled in Terence's computer or somebody else's, then there's not much space left for something truly new to emerge. My view is that the complexity of Terence's novelty wave, as a concept of the quality of time, might develop as the universe develops, not according to some simple algorithm but in a way that's truly unforeseeable.

TERENCE: This would arise out of its resonance with its own past. It's in the realm of resonances that this very difficult-to-quantify complexification accumulates.

RUPERT: My problem is that I think spontaneity and creativity come first. All these attempts to construct models of what's going on will be inadequate to encompass the phenomena.

TERENCE: All models are provisional, and that's what preserves the open-endedness of what they are modeling. The great intellectual and emotional change accompanying the paradigm shift will be in people's ability to accept not having full explanations; they will understand that

the depth of the mystery exceeds explanations. Models are always provisional, always made of string and sealing wax. There is a part of reality
that has been referred to as the value-dark dimension, about which
nothing can be absolutely known. Chaos is in there somewhere. The task
of human becoming, and of mathematics and our intellectual tools, is
to cast light into this dimension. This gives us a sense of discovery and
meaning, but it can never reduce the dimension to absolutes. Eventually,
there is a domain in the value-dark dimension that is value-dark in *principle*. The physical analogy that we have for this is a black hole, from
which no information escapes. We have to build into our theories these
kinds of trap doors and escapes so that we don't get caught inside
another illusion.

RUPERT: The attempt to tame chaos by having mathematical models of
it is a modern version on a rather abstract plane of the old myth of the
solar hero conquering the sea monster, the dragon of the deep, the
serpent of chaos. In the Christian version, Saint Michael or Saint George
is the shining hero piercing the darkness of the dragon. Ralph said that
we can model some of these chaotic processes but that there might be a
small, medium, or large residuum. Let's assume it's large. Then we build
more models to model some of the residuum. From there, it would be
easy to slide into the familiar statement, "Well, in principle, it should be
possible to model it all." This is the modern version of the idea that
somehow, in principle, the whole of reality can be engulfed within some
kind of mathematical model. In other words, the world soul is in some
sense subject to the supreme mathematical mind, which is superior to,
transcendent of, and prior to the whole natural world. This slides over
into a kind of metaphysics that's very traditional among mathematicians.

RALPH: We've got to have a talk about this, Rupert. You persist in thinking that mathematics hasn't advanced since the time of Plato. Your
objection to models for chaos is based on an inappropriate view of
Platonic ideals. The new models for chaos are coming into existence by
a process of evolution and discovery. I don't think that Plato, or anyone
before the invention of the computer, could even have imagined them.
This discovery process is part of the evolutionary, creative aspect of the
world soul, which we can think of now in two layers. One layer is the

world of matter and energy in which there is the discovery of new fields in evolution. The other layer is the mental world, which includes verbal descriptions and mathematical models.

Apparently, these two levels are in a process of coevolution. If, as in figure 3, there is a Gaian unconscious supplying new forms out of chaos as raw material for evolution, then either it is supplying both levels, the mental and the physical, or else the mental and the physical are two separate levels to the Gaian unconscious. Probably, there's a connection from the Gaian unconscious to each of the two coevolving levels; this three-way connection could be modeled as a triangle. The idea that whatever comes into consciousness has a mathematical model doesn't conflict with creation, because we can't assume that all mathematical models already exist. There's the same infinite possibility of discovery, invention, and novelty in this mental plane as in the material. Why not? What's the difference?

Rupert, you seem to think that one or the other of these has to have precedence over the other—that either the mathematical model is abstracted from material observation or the material world is concretizing or condensing around mathematical models. This is unnecessary. It may be just a process of coevolution. Sometimes the left foot leads and sometimes the right.

RUPERT: Fine. That's a wonderful explanation. However, whereas you're a mathematician yet not a Platonist, Terence is much more Platonic. He thinks the novelty wave is coming from some kind of higher realm and somehow underlies the behind-the-scenes mechanics of the cosmos.

TERENCE: I'm not sure whether this novelty wave is simply a mathematical description of an enzymatically mediated process on the surface of the Earth or, as you indicated, whether it can be raised to the level of a higher principle. It is determined, but only in a fairly weak way that sets the schedule of events without announcing what the events will be. I'm troubled by my Platonism. I realize that certain naive assumptions haunt it. I don't want these forms to be eternal. I want them to somehow arise internally out of the ongoing process of the world. I haven't quite figured out how to get these ducks all in a row.

RALPH: Before the computer revolution, there might have been some question as to the preexistence of all mathematical forms. But now, there's no question. Mathematics is a world of its own, a landscape with hills and valleys, and much of this terrain has not been explored. Some features have been identified by various travelers who have come back and given their reports. There is a process of discovery of the already existing landscape as well as the modification and evolution of that landscape through the interaction of human consciousness. Although it's one, two, three here, and one, two, three everywhere, we could still land on a foreign planet and find that they had explored, excavated, and modified a region of the mathematical landscape never visited by us.

TERENCE: So little of mathematics has to do with numbers. It's basically paying attention to the rules operating among defined sets of objects. I think that once you figure out what mathematics is, every single one of us could invent a new branch of it.

RALPH: I don't believe that mathematics is the result of creative activity on the part of people, that it's been invented. It hangs together with an integrity that is beyond the capabilities of the short history of human consciousness.

TERENCE: Does it hang together? Or is it an archipelago of islands? Do the people who are doing advanced number theory have anything whatsoever to say to the super algebraists who don't have anything to say to the fractal people?

RALPH: I think it's a single landscape, all tied together in complete integrity.

TERENCE: A wild-eyed claim!

RALPH: It doesn't matter too much. The varied mathematical theories are just impressions of travelers having returned from a distant land. I think it's time for us to face the soul of the world. We know from reports in the Library of Congress that other people have seen in their travels a different part of the world soul. Mathematics here, sensory experience there, Babylonian history here, chaos of the unconscious providing novelty there, and so on. How do we put it all together? What is the relationship between the existence of the world soul and the human imagination? The mathematical landscape was extensively traveled, mapped, experienced, and used in the biological world long before the human species evolved. What is the role of mathematics in the coevolution of the world soul and human consciousness?

RUPERT: One can deduce several things about the soul of the world. One is that it contains qualities as well as quantities. The world we actually experience is full of colors, sounds, smells, and other qualities known to us through our senses. The procedure of science since the seventeenth century has been to ignore sensory qualities and to consider only what were called the primary qualities of substances, namely, their weight, position, momentum, and so on. These could be assigned numbers and treated mathematically. Reality was treated as colorless, tasteless, soundless, and odorless. It was abstract, objective, and mathematical. Qualities known through our senses had no objective existence outside of the mind of the subjective observer.

It seems to me that the imagination of the world soul is going to work, not just in terms of numbers and mathematics, but also in terms of qualities. It's likely to contain all possible tastes, smells, colors, and other qualities that exist in the world, as well as the experience and imagination of these qualities.

TERENCE: Heaven.

RUPERT: No, it's not heaven. It's the soul of the world. The soul contains not only everything that's in the world but also the imagination that has given rise to everything in the world. And this imagination is continually active, giving rise to new forms and possibilities.

RALPH: Does that mean the soul of the world is not evolving but is already complete?

RUPERT: To answer that, let me look first at a more conventional view of the soul of the world. Theoretical physicists are currently trying to conceive of a unified field of everything. This was Einstein's goal, but he couldn't reach it because he tried to do it by mathematics alone. This problem is now seen in the context of the Big Bang cosmology.

When the whole physical universe is cranked by calculation right back to the first few jiffies, the temperature rockets up to billions of degrees centigrade and everything changes. Things don't behave the same way they behave in the present. They become more symmetrical. When the universe is cranked back even further, it arrives at a state of primal unity, the primal field of nature. According to superstring theory, this field has nine dimensions of space and one of time. As the universe develops and expands, symmetries break, and the fields of nature, such as the electromagnetic and the gravitational fields, precipitate out.

All forms and patterns of things that develop in the world have their own organizing fields, and all are ultimately derived from the primal unified field, which remains the all-encompassing field of the world. The world field, since it contains everything within it, necessarily has an evolutionary quality because it embraces everything that happens in the evolving cosmos. This is the conception toward which modern cosmological speculation is pointing.

In my own view, the world field has a memory of everything that's happened within it already. The cosmic imagination involves ongoing memory in a world whose physical body is shaped by the habits of nature. Thinking of the world soul in terms of a world field still has a kind of black-and-white, mathematical, abstract quality to it. What I want to express is that any reasonable conception of the world soul would have to recognize the existence of colors, tastes, smells, sounds, and other qualities.

TERENCE: We use the word *soul* for the world soul because we sense an analogy with the soul of the individual. If you begin to carry forward that analogy, you get into some fairly astonishing places. We imagine the soul to be a nonlocalizable, nonmaterial essence that survives death, a higher dimensional form erected through the process of life that when the body dies is released into a higher dimension that is its source and home.

Pursuing this analogy, is the world soul an invisible, unseen, organismic structure that has been erected through the evolution of life on this plane? Is the destiny of the world soul incomplete until it severs itself from the matrix that created it? Is the global crisis and inner searching and turmoil of our time the dawning realization that we are actually facing the death of the world soul, meaning its severance from the dimensions that allowed it to accrete and form?

RALPH: I must say I'm feeling very uncomfortable with this discussion, and I'm astonished to find myself sitting here accusing you guys, of all people, of thinking too small. I find the whole idea that the world's soul is confined in a space/time continuum of four or ten dimensions extremely claustrophobic, as well as the idea that the world soul had no chance of existence until the Big Bang provided matter and energy or something. I'm even doubtful about the Big Bang. But even assuming that it occurred, I think the whole idea of soul suggests the aspiration of eternity for consciousness and unconsciousness or for some ultimate essence of the life experience. It could be that solar systems come and go, that universes come and go, and that the world soul, as it were, incarnates in one universe after another. After a Big Bang, there may follow a collapse, followed by another Big Bang. The idea of the world soul coming to an end once and for all because of a nuclear winter or something is very confining.

TERENCE: I'm not suggesting that it ceases to exist. I'm suggesting that it's liberated into another dimension.

RUPERT: Ralph, yours is the Hindu model, and Terence's is the Christian.

RALPH: Perhaps. I'm talking about an existence of the world soul that is beyond space and time. Space and time are illusions having to do with the severe restrictions of incarnation in these chimpanzee bodies.

RUPERT: This is pure Hinduism. It is a Hindu doctrine that the soul comes from a realm beyond space and time and is incarnated in a body. When its body dies, it is reincarnated in another one. The soul's true destiny is not in the body. It remains in touch with the source from which it has come, which is far greater than any of its embodied existences.

One possible traditional model is that the universe will die, and then the cosmic soul will be reincarnated. I contrast it with Terence's model, which is the Christian eschatological model in which the universe reaches a culmination. Theidea is that the whole of creation, the entire universe, is groaning in travail for a new order to be born. This is the idea of a new creation—not just of humanity, and not just of this Earth, but of the whole cosmos.

TERENCE: Out of Bios.

RUPERT: Out of Bios comes a totally new order of existence. This corresponds roughly to your notion of the embodiment of the world soul coming to an end—presumably not just on this Earth, but in the entire cosmos. This is the most extreme version of Judeo-Christian eschatological thought.

TERENCE: Devil, you say. This is most extreme. Out of devotion to my theory, I give assent to it. However, my personal notion of the world soul is not as metaphysical as either of yours. I'm not really into it as God Almighty. I'm into the world soul as the largest and smartest creature imaginable. It isn't the god who hung stars like lamps in heaven, and it isn't the force that spins the galaxies on their axes. The world soul is something that has arisen out of biology. It's an organism within the universe of space and time, but it didn't make the universe. It's an inhabitant of it, but on a scale that makes us mere atoms within its form.

RALPH: I think we're all involved in a kind of compromise. What we need here is a Twelve-Step group or something. We have to reprogram ourselves out of our childhood conditioning as the Hindu, the Judeo-Christian, and the Scientist. Our work, thought, talk, and relationship are very much inspired by our individual travels through the spirit, where we have seen and felt the largeness of the world soul. When we discuss it or bring it down into language or relate it to ordinary reality, there is a tendency toward conservatism, which ends up looking like anthropocentrism. We compress what we've experienced on a grand scale down to the human scale and relate it too much to human consciousness and human history.

TERENCE: When I look at human history, I see the accumulation of a sense of urgency long before anyone started worrying about ecocide or population. It's almost as though the world soul is the thing that wants to live and, sensing instability, it is trying to build a lifeboat out of the clumsy material of protoplasm.

RALPH: This is like fighting an infection. The Gaian mind may be faltering due to a bad habit. Incarnation is addictive, and the reason there is an infection out of control on this planet is because of this bad habit.

TERENCE: The world soul may actually sense the finite life of the sun, and it may be trying to build a lifeboat for itself to cross to another star. How in the world can you cross to another star when the only material available is protoplasm? Well, it may take fifty million years, but there are strategies. They have to do with genetic languages, and with developing a creature who deals with matter through abstraction and analysis, eventually creating technology. This is all an enzymatically mediated process, a plan in the mind of the world soul to survive.

RALPH: In our experience of the divine logos, isn't there the feeling that we've already gone beyond the physical plane of protoplasm? Is this not already a kind of star travel?

TERENCE: Yes, but then why this increasing urgency, century after century? For fifteen thousand years, there has been increasing anxiety and the following of increasingly irrational chreodes. Only if there is a problem with the stability of the environment do the last ten thousand years of human history make any sense. This problem has created history as an evacuation, a frantic project to find a way out. That's why things have been allowed to tear loose, to poison the oceans, to strip the continents. The world soul, I think, is in communication with us in the culminating moment of human history. Everything is being scripted for a purpose, and toward an end unglimpsed by us but tied up with the survival of everything.

   If it were seriously important to attain star flight at all costs because the biosphere is in trouble, a fruitful approach would be superminiaturization. In other words, we would need to find a way to turn people into spores and then seed these spores throughout the galaxy, relying on light pressure and gravitational convection to distribute them. At the rate of percolation of matter through the galaxy, spores released from a single planet could penetrate the entire galaxy in about forty-five million years, which on the scale of the life of the universe is not long at all. I think what is called for is a retooling of the human form.

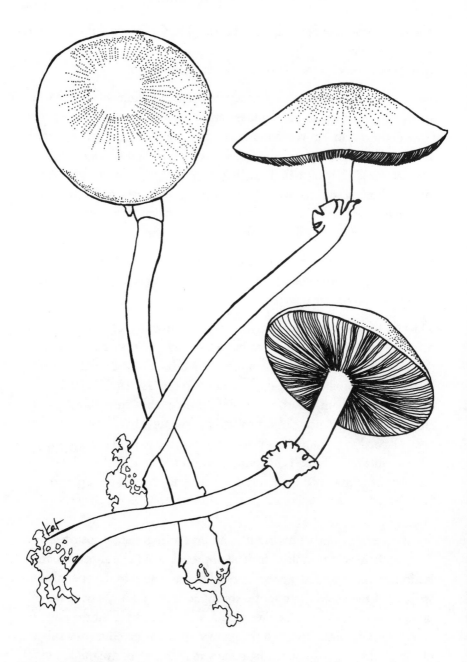

Figure 6. *Three perspectives on* Stropharia cubensis, *a psilocybin-containing mushroom. Kathleen Harrison McKenna* © *1991.*

RALPH: We're going to send termites out into space?

TERENCE: Mushrooms! Human mushrooms. We know that psilocybin is closely related to serotonin. Serotonin makes the brain functions of the mental universe possible for the mushroom.

If you think about the mushroom, it is perfectly engineered for truly long-duration survival and adaptation. Look how lightly it touches matter. Its mycelium is simply a cobweb in the soil of any planet, and yet it synapses upon itself and is full of neurotransmitter-like psychedelic compounds. It's like a thinking brain, yet it condenses itself down into a thing three microns across, of which several million per minute can be shed by a single carpophore. Spores are perfectly designed to travel in space. They can endure extremes of temperature. Their color reflects ultraviolet radiation. The surfaces of spores are composed of the most radiation-impervious organic materials known.

This is an example of how an abstract notion like the world soul can penetrate the upper levels of the world of biology and organisms.

RUPERT: Insofar as the mushroom and the human psyche have had a symbiotic relationship, the mushroom-induced experiences in human consciousness are in the morphic field of this symbiotic relationship. Therefore, these experiences could be carried by the mushroom. The mushroom spores would have to germinate somewhere, giving rise to mushrooms on another planet. Then, when conscious organisms ate the mushrooms, they would gain access telepathically to this whole realm of the human psyche.

TERENCE: It's more than a symbiosis. Perhaps we are going to be downloaded, or uploaded. If we can find a way to download ourselves into the mushrooms, then, when the planet explodes, it'll be a free tailwind to our tour bus, you see.

RUPERT: I think we may have already downloaded ourselves into the mushrooms.

RALPH: The world soul as mycelium: the pattern that connects . . .

In fact, the whole Earth may be chemically regulated through very small molecules, aromatic compounds that are byproducts of the metabolism of various species but that percolate out through the environment and set up the ambience in which a lot of animal and plant business is done.
   —Terence McKenna

Just as the electromagnetic field is an interface between the matter fields and the mental, psychic, and morphic aspects of ourselves as human beings, so the electromagnetic field could be playing a similar role in the mental structure of the soul of the world.
   —Rupert Sheldrake

We have in our individual consciousness a particular affinity with the electromagnetic field: electromagnetic perception, reception, and so on, as epitomized by vision. The easiest thing to affect by the phenomena of mind over matter should be the electromagnetic field.
   —Ralph Abraham

5

LIGHT AND VISION

RUPERT: I've been thinking about the connection between physical light and the light of consciousness, the light of reason, and the light of vision. It's not enough to say that one kind of light is physical and the others are metaphorical. In some sense they must be aspects of each other.

The key may be in the connection between light and vision. How much do we understand about the nature of vision? Science doesn't tell us much. It tells us that light moves from the thing we see, goes through the eye, forms an inverted image on the retina, and causes patterns of electrical and chemical activity to take place in the optic nerves and cerebral cortex. Then, somehow, what we're seeing seems to spring up in a totally unexplained way as a subjective image. This image is somewhere inside the brain, yet it is subjectively experienced as outside the body. If I'm looking at you, Terence, the light rays come into my eye, and then I have a subjective image of you that is, according to the standard theory, an electrochemical pattern in my optical cortex. This seems to me an extremely peculiar theory of vision, because it locates the visual world we experience inside the brain and not around us where it seems to be.

When I look at you, my image of you is interpreted by me; it's a mental construct. I think this mental construct may not be inside the brain but right where you are, namely, outside me, where my image of you seems to be. The conventional idea that the image is inside my brain does not correspond with my actual experience. It's just a theory, but a theory of remarkable hallucinatory power because we so easily forget it's merely a theory.

If in the process of vision there's an outward projection of images as well as an inward movement of light, if I'm not just playing with words, then there must be something moving out as well as light moving in. If so, people or things might be affected just by being looked at.

The idea that something goes out from the eyes is a very old and traditional view of vision. It was present among the pre-Socratics and is found in implicit form all over the world in the fear and practices associated with the "evil eye." These practices are indeed supposed to involve the outward movement of influences from the eye to the thing or person being looked at. There's also an enormous folklore in Western culture about the sense of being stared at, the feeling people have when they think they're being looked at, for example, from behind.

There's not been much empirical research on the sense of being stared at: three published papers in a hundred years. It's a subject that parapsychologists have ignored as well as psychologists. It could be, oddly enough, the biggest blind spot in our view of the world, because it could hold the key to an entirely new understanding of the relationship between mind and matter, or spirit and body.

Let's assume for the purpose of discussion that it can be established empirically that there is indeed such a thing as the sense of being stared at. Some influence passing out through the eyes can be detected empirically. What kind of influence could this possibly be? What kind of influence could be moving outward through the eyes in the opposite direction to the incoming light?

There are two possible ways of explaining such an influence. First, this outward movement could be in a kind of mental field that is somehow over and above the electromagnetic field. The electromagnetic field sets off electrochemical changes in the brain, and somehow the mental field organizes and meshes in and relates to it but is not itself part of it. The mental field then projects outward, placing an image where the person or object being viewed is actually located. Such fields would extend all around us and be filled with our sensory experience of the world.

In the second model, a more economical one, the outward projection process takes place by a reverse movement through the light that is coming into the eyes. In other words, when a photon of light comes in, it corresponds to an antiparticle moving out, and these outward-moving influences move along the exact track of the photons. These particles are then associated with vision, perception, comprehension, and all subjective experience of an object. They're the grok wave, if you like. Since, physically speaking, from the point of view of a photon, no time elapses as it travels, the connection between the source and the place where it arrives, between subject and object, is instantaneous. In this way vision may be very closely related to light.

It would be crude to say that the antiparticle of the photon is the visi-on—the particle of sight—but this concept may be the missing link between vision and light. It may simply be that the photon is in some sense reversible, and that the electromagnetic field is in some sense the field of vision as well.

TERENCE: There are a number of questions to be asked here. If the visi-on were simply the antiparticle of the photon, the phenomenon of light pressure should not exist. Since this phenomenon does exist and is well studied, the visi-on must be more esoteric then an antiphoton.

RUPERT: This is no problem at all. The photon is a particle with physical properties in the physical realm and therefore exerts physical pressure. The visi-on has to do with conscious experience, with properties of mind. It's moving in the opposite direction, and the sensing of being stared at would be a sensing of those particles or waves impinging. There would be a kind of pressure in that direction of the psychic kind.

TERENCE: A more elegant way to describe what you're saying would be to call this reverse wave phenomenon a quality of the photon itself. From the point of view of the photon, the travel time to and from its destination is zero. Likewise, travel time from destination back to origin is zero. Why not simply take a page from superstring theory and visualize the photon as a kind of particle that is stretched in one dimension? It is present at its origin and its destination simultaneously and thus able to impart information at a distance.

RALPH: As hypotheses go, this is the big bag full. Certainly if it were established that no effect could be produced in a person by looking at them from behind, this discussion would be less interesting. There would still remain some serious outstanding questions about the morphogenetic fields proposed as the memory banks of the species in Rupert's books.

RUPERT: There would be questions as well about the relation of mind and matter and the nature of vision and light.

RALPH: Nevertheless, it doesn't seem very useful to make the assumption that such an effect has been established. If it were, it would be the first of all the various so-called paranormal phenomena of recent decades to be validated and accepted. That would mean a lot of paradigms would be shifted, and we would also be seeking explanations for telepathy, remote viewing, clairvoyance, and other effects of vision at a distance.

RUPERT: Second sight.

TERENCE: I don't think it would automatically follow, if this effect were confirmed, that people would think it had anything to do with eyes or light at all. If we establish that attention can be felt across space, this would be an establishment of telepathy. The felt sensation would not be due to the fact that I'm looking at the back of someone's head, but that I'm focusing my attention, which would be most effective when I really bore in. The output of the optical system does not increase; it's the output of the mental act of concentration that does.

RALPH: If we agree, then, that science would be stood on its head by the discovery of one paranormal thing, which would make all paranormal occurrences fair game, then we can assume that field theories would expand considerably. Mathematical models for the morphogenetic field would abound. Let's say we did have an experimental result in the laboratory demonstrating that a person can get someone else's attention by staring at them, or by boring in, as Terence says. Would it really be natural then to propose the electromagnetic field as an intermediary for that influence? Or would we rather propose another field as a conceptual model for the observed phenomenon?

Since ancient times, we've thought of mental, physical, and spiritual phenomena as operating in different planes. The planes used metaphorically by the ancients of Greece and the *rishis* (seers) of India are more or less what we're calling fields, and these thinkers found it useful to separate these fields.

The electromagnetic field is physical—at least I think of it as physical. I like the idea of a separate mental field. The mind somehow follows the eye and extends itself so as to actually engulf the object being viewed, to know it through intimate touch. Cognition is then a kind of engulfing, like eating. This motion is visualized in the mental plane and therefore belongs to a different field. I feel a bias toward this view.

On the other hand, if you could show that this telepathic transmission from one person to another can be modulated electromagnetically by a magnet or filter, that would strengthen the argument for the actual proximity of the two separate planes without, even then, identifying them.

RUPERT: There's a part of me that thinks the separate field idea is more attractive, but I've been leaning over backward to see whether we can come to a new understanding of the electromagnetic field in which the connection between vision and light is a very close one. Even if there is another field involved, it must be in intimate resonance with the electromagnetic field. There's no doubt that changes in the brain are largely changes in electromagnetic patterns, and there's scope for resonance there, but the mental field may also resonate with the light that's coming into our eyes. If this mental field is resonating with the electromagnetic field of light, then indeed it will connect us through the light to the objects we are seeing. Anything that resonates with the light that's in between us will directly connect us via the light.

RALPH: I'm more comfortable with the wave rather than the particle metaphor. Let's just think of waves. Here I am looking at waves on the ocean, and I see that there's a rock out there. As the waves pass the rock, their shape is changed: there is a hologram of the rock within the wave that comes forward and crashes on the beach. Then there's a reflected wave that goes back. I think the electromagnetic field, as physicists view it, is something very much like this. Its mathematical model is a wave equation. This seems to be a suitable medium for influence to go both ways. But somehow I don't see it as having a rich enough structure to model all mental processes.

RUPERT: I'm only modeling vision, so far.

RALPH: The visual part of the mental field may be a very thin slice of the morphogenetic field that is in very close resonance with the electromagnetic field.

RUPERT: I think that if we take the idea of interfacing planes—the old idea of different levels—and think of the planes as fields, then we do actually have a series of stratified levels.

First there are the quantum matter fields, which have to do with the strong and weak nuclear forces in atoms and which determine the shape and structural properties of atoms and molecules. They only work at very short ranges.

Then there is the electromagnetic field, which is an organizing field of more complex structures. The electromagnetic field actually holds together atoms, molecules, crystals, and everything else. One could say that the electromagnetic field is associated with the morphic fields of molecules and crystals.

At the level of plants, there's a morphogenetic field of vegetative growth that somehow interfaces with the electromagnetic field. In animals, over and above the morphogenetic fields are the fields of instincts and movements; they organize and coordinate the activities of the nervous system.

There are hierarchically higher planes above these, such as perceptual fields and fields of higher level understanding. There may be planes of fields that are like the levels you spoke of, Ralph, but in a nested hierarchy. The gravitational field embraces all; it's the universal field.

RALPH: The electromagnetic field, a constituent component of the morphogenetic field or the world soul, should be utilized economically to carry as much of the burden of explanation as possible. Certainly all of the morphogenetic phenomena of crystals and so on are intimately connected by resonance with the electromagnetic field.

TERENCE: There are not only electromagnetic fields but chemical fields. I think that pheromones are vastly underrated for their organizing power in biology and social systems. In fact, the whole Earth may be chemically regulated through very small molecules, aromatic compounds that are byproducts of the metabolism of various species but that percolate out through the environment and set up the ambience in which a lot of animal and plant business is done. Easily volatilized low-molecular-weight compounds are probably behind a lot of the mechanisms for the self-regulation of nature. If materialists can seriously argue that the progressive ease through time of crystallizing new compounds has to do with seed crystals moving around from laboratory to laboratory on chemists' beards, then they will certainly be in agreement that the percolation rates of nature are effective enough to move these control- and message-bearing chemicals around everywhere.

RALPH: This idea is very supportive of Rupert's economy move, because the olfactory bulb is nothing but a transducer of information from the chemical field to the electromagnetic field. Just a small number of molecules of a pheromone are enough to excite an identifiable electromagnetic wave across the bulb, which is then identified by some kind of associative memory living primarily in the electromagnetic activity of the brain. This coupling shows that the electromagnetic field is an intermediary between the chemical field and the mental field.

TERENCE: The chemical field is simply a higher-order manifestation of the electromagnetic field, because most of these volatile compounds have very electronically active ring structures.

RUPERT: It's a resonance phenomenon.

TERENCE: It is charge transfer and resonance, and doubtless bioelectronic activity of other types. The most electronically active molecules are the drugs, the pheromones, the growth regulators, and so forth.

RUPERT: These principles could also apply to hearing. When I see you, you are localized somewhere outside me, where you are. If I hear you, your sounds are also localized outside me. I don't hear sounds as if they're arising inside my auditory cortex. I hear them as if they're rising around me in three-dimensional space, and I can locate which direction they're coming from.

This means we are not only surrounded by a visual perceptual field that spreads out from us and fills the space of our perception, but we are also surrounded by an auditory perceptual field. We are surrounded by an ocean of fields.

RALPH: The ocean has infinite structure and complexity but nevertheless could never function as a brain. The brain is, in the neurophysiology it presents to the experimentalist, certainly much simpler than the mind. The brain cannot function through the electromagnetic field alone, even though all of its effects and patterns are manifest in the electromagnetic field. Its structure is much richer than the electromagnetic field, which can't control all these patterns without the brain's complex structure of cortices, intercortical cells, chemical messengers, and ion channels.

Modeling the brain requires much more mathematical structure than does modeling the electromagnetic field. The brain is much closer to the physical universe than to the mental universe, so I think the electromagnetic field is too thin to occupy more than a fraction of the entire structure of the field that carries recognition, memory, the ability to serve in tennis and learn a new language and recognize haiku, and so on.

RUPERT: I think that somehow the brain plays an interface role between the chemical and morphic and mental realms. The question that arises is, How does the electromagnetic field interface with the quantum mechanical fields that hold together the structures of atomic nuclei and electrons in their orbits? These structures are actually maintained by fields that in a sense are stronger than the electromagnetic field, for they resist it in such a way that the positively charged protons bound together in the nucleus do not fly apart through mutual repulsion, and the negatively charged electrons do not plunge into the nucleus. The electromagnetic field works around these matter fields as a more subtle field.

RALPH: Perhaps the quantum mechanical field, not the electromagnetic field, is the intermediary between the physical and mental planes.

TERENCE: But, Ralph, if you feel that the electromagnetic is inadequate to what Rupert is asking of it, then you must be equally skeptical of the morphogenetic field.

RALPH: No, I'm not skeptical. I think that what we're trying to do, through the revision of our actual life experience, is make up a model for some of the paranormal phenomena we've experienced that scientists prefer to totally ignore. The electromagnetic field and the history of its modeling, its hermeneutics, is an excellent case to imitate. What we are trying to do is fashion a field concept around these phenomena. It does seem very attractive to think of the electromagnetic field as some kind of favored intermediary among all the physical fields.

Perhaps the mental field will end up with a mathematical model that is field theoretic, multidimensional, and coupled only to the electromagnetic field, which is coupled to all the other physical fields.

This is a combination model, a sandwich model, that might be successful in explaining perception, cognition, and the idiosyncrasies of time. There might eventually be a general relativity theory for the mental field that would explain clairvoyance and so on. What we're talking about now is the struggle to envision the coupling between the mental field and the electromagnetic field, including wave metaphors, particle metaphors, and so on.

RUPERT: That's right. My preferred model has always been one in which perception and mental activity have a kind of fieldlike structure; it would be a morphic field of some kind.

RALPH: We also ought to think about the possibility that these effects, like the sense of being stared at, will not be confirmed in laboratories. This is already the experience of many experimental efforts over the years to confirm the so-called paranormal. Where there is evidence, it always seems to be just the slightest bulge of the curve to the right or left of absolute insignificance.

These paranormal phenomena, which we want to capture in a model, don't seem to be very robust. They come and go, perhaps due to the fact that they belong to the mental field and are very subject to noise in that field. For whatever reason, they are very difficult to confirm, and I'm thinking of that as being a kind of evidence in itself. On the one hand, there is the very widespread impression that these things exist, and on the other hand, there is the impossibility of confirming them in the experimental paradigm of modern science. Somehow this suggests to me the necessity for a thick model, a richer field in which to try to do the modeling, where there are chaotic attractors everywhere and no homeostasis or anything like that.

RUPERT: The reason I wanted to discuss this in the first place is that I actually don't know what to think. Taking light and vision to be two aspects of the same phenomenon leads us into a whole other area: the seemingly metaphorical meaning of light in the context of the "light of

consciousness." For example, when we dream, we see things in a kind of
light. This light illuminates psychedelic visions, dreams, daydreams, and
all visual imagery that occurs with our eyes closed. There's some sense
in which our imagination, our image-making faculty, is self-luminous.

In the case of vision in normal physical light, the light comes first
and vision comes second. If you shut the light off, you can't see. But in
the visionary sense, vision itself may generate light, at least subjectively.
Visions are self-luminous. If someone has visions enough, according
to religious traditions, they start developing halos and their bodies
become luminous.

The point I'm trying to make is that if physical light has conscious
vision associated with it, then the reverse may also be true: conscious
imagery may have light associated with it.

TERENCE: Tryptamine hallucinogens certainly fill the head with light.
Serotonin, their near relative in brain chemistry, is transduced to melatonin by a light-mediated reaction. In other words, light actually enters
through the eyes and follows a part of the visual pathway that branches
off and goes to the pineal gland, where photons work a chemical change
on serotonin and turn it into melatonin. These compounds are near
relatives of the tryptamines, which are the psychoactive compounds
occurring in psychedelic mushrooms. All this is going on in the pineal,
and it's all light-driven chemistry.

RALPH: Nevertheless, it seems that this kind of vision has nothing to do
with the electromagnetic field. Even though these neurotransmitters are
very photosensitive, the fact is that in the dark people have very bright
visions. Neurochemical activity in the visual cortex or somewhere else
mimics the effect of photons falling on the retina. In the alternation
back and forth between electromagnetic and chemical waves on neuro-
physiological levels, the psychedelics take over at some point and supply
what appears to be the result of a previous train of several reversals.
In the illumination on that level, the photons are replaced by this
other messenger.

It might be useful to think about the habits of electromagnetic nature
and the behavior of the electromagnetic field as evolving according to the
habits of the morphogenetic field. This would explain the resonance

between them: as above, so below. The m-field creates the em-field to do its bidding.

RUPERT: Just as the electromagnetic field is an interface between the matter fields and the mental, psychic, and morphic aspects of ourselves as human beings, so the electromagnetic field could be playing a similar role in the mental structure of the soul of the world. If there is a world soul that permeates the entire cosmos, its bodily level may be expressed primarily through the gravitational field while its mental level may be expressed through some kind of interface with the electromagnetic field. This would be a perfect medium for the world soul's omniscience and for divine omniscience through the world soul. Everything that happens affects the electromagnetic field; its holographic reality at any moment is in exact correspondence with what's happening. The universal electromagnetic field is the interface of the world soul with the physical planes of reality.

RALPH: I like the idea of the electromagnetic field being an ideal intermediary positioned in a hierarchy of fields, much like the position of our individual consciousness in the hierarchy of consciousness of the world soul. We have in our individual consciousness a particular affinity with the electromagnetic field: electromagnetic perception, reception, and so on as epitomized by vision. The easiest thing to affect by the phenomena of mind over matter should be the electromagnetic field.

RUPERT: There are a lot of people who think the universe is conscious or that the soul of the world in some sense is perceiving what's going on. There are many theological traditions of divine omniscience. By definition, any theology of divine omniscience requires the divine mind to know everything. Knowing everything would include knowing all the properties and states of the electromagnetic and gravitational fields, so these would be essential aspects of divine omniscience.

I've found that when people think about divine omniscience, they treat it as an entirely miraculous process, totally disconnected from any kind of physical reality. However, divine omniscience must involve knowing from within—being within all things. Therefore, cosmic omniscience must pervade the electromagnetic field and all the fields of nature. This is like Newton's notion that the medium of divine omniscience was Absolute Space, which he called the sensorium of God.

TERENCE: Why is divine omniscience a necessary concept? Can't the universe get along just being partially aware of what's going on?

RUPERT: I think that it's intriguing to consider models of reality in which there is a sense of knowing associated with the cosmos.

TERENCE: We all have a sense of knowing, but we're not omniscient.

RUPERT: There are two possible models. One is the standard model of secular humanism, which postulates that our minds are the most advanced in the universe. According to this model, the rest of the universe is essentially unconscious. Living organisms crawled out of the primal broth in an inanimate universe and, through the miracles of random mutation and neo-Darwinian natural selection, gave rise to organisms such as ourselves with complex nervous systems that have the subjective correlate of consciousness. Human consciousness emerged out of the darkness of inanimate nature and is the highest consciousness that exists, although it's conceivable that intelligent beings have evolved on a few other planets as well.

The more traditional model derives human consciousness from a much larger consciousness that pervades the cosmos, the Earth, and the whole of life on Earth. In this model, our consciousness has come about by a kind of diminution or descent of some higher kind of consciousness rather than by an ascent from lower, ultimately unconscious matter. We're a reduced form, a self-contracted version, of a higher consciousness rather than an inflated version of a lower, animal consciousness.

I find it more reasonable to suppose that our minds are in touch with larger minds, and that in many ways they are shaped by larger mental systems of societies and cultures, ecosystems, Gaia, the galaxy, the entire cosmos, and perhaps by a cosmic mind beyond that.

TERENCE: I follow you as far as a Gaian mind because it seems a biological object, not a theological premise. To hypothesize a mind of the whole universe seems unnecessary and unlikely to be encountered.

RALPH: Suppose there was another Gaia, another inhabited planet. It would have its own Gaian mind.

TERENCE: The inhabitants would be citizens of the universe, even as we are.

RALPH: Between our Gaian mind and theirs, a dialogue might be in progress.

TERENCE: Nothing of this suggests theology.

RALPH: You're projecting theology onto it in association with the phrase "divine omniscience." This is usually associated with Yahweh, the one God of the Hebrews, and his divine omniscience. It's unnecessary to make that association. As you've gone as far as the Gaian mind, you can certainly go further. There is a hierarchy of worlds in the universe.

TERENCE: I'm not convinced that the hierarchy is minded at every level.

RALPH: That's the question. Does Jupiter have a Jovian mind?

TERENCE: Does the Solar System have a mind?

RUPERT: In any holistic model of reality, it seems entirely natural to suppose that Gaia has a kind of mind, and that the Gaian mind is embedded in the Solar System mind, and the Solar System mind embedded in the galactic mind. These higher levels of mind and consciousness, which may be hard for us to conceive of, seem likely to exist by a simple logical argument.

TERENCE: But they may not necessarily exist in a hierarchical order. After all, chipmunks are not small portions of whales.

RUPERT: No, no, no, but chipmunks are not inside whales. Chipmunks are inside terrestrial ecosystems. Whales are in different ecosystems, and both of these, the oceanic and the terrestrial ecosystems are part of Gaia. The hierarchy is of more inclusive units; it is a nested hierarchy. The Solar System is not on the same level as the Earth. It's a higher level of organization of which the Earth is part. The galaxy is a higher level of which the Solar System is part.

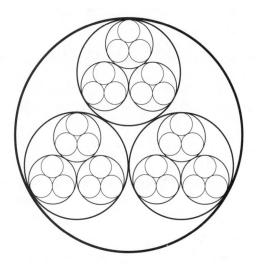

Figure 7. Nested Hierarchy. *Diagrammatic representation of a nested hierarchy, which could represent subatomic particles in atoms, molecules, or crystals, for example; cells in tissues, organs, or organisms; or planets in solar systems, galaxies, or galactic clusters.*

RALPH: If you accept a Gaian mind, do you think there's a similar kind of mind associated with Jupiter?

TERENCE: Yes. Planets are like animals.

RALPH: What about the solar mind?

TERENCE: Other places in the Solar System seem potentially more open to the support of recognizable forms of life. The oceans of Europa, for instance.

RALPH: Does something need recognizable life to have a mind? In the Gaian mind, not only the so-called living beings have mind. The ecosystem has a mind.

TERENCE: But you don't want to define mind so broadly that it's no longer recognizable as what we say when we ordinarily use the word.

RALPH: That's a problem, I agree. But all these different components of the cosmos appear to fit exactly our model of thinking. To ask if some-

Figure 8. Eye of Horus. *The radiant "Eye of Horus" on the Great Seal of the United States, as shown on every dollar bill.*

thing is conscious or unconscious is a different question that very much complicates the issue.

TERENCE: I'm definitely into the notion that if you can't have an I-thou relationship with it, it's fairly uninteresting to call it a mind.

RALPH: Do you think then that the Gaian mind would be dead if the human species became extinct?

TERENCE: Not any more than someone would be dead if an acquaintance dies. I communicate with the Gaian mind.

RALPH: Then why can't Jupiter have a Jovian mind without the necessity of microbes?

TERENCE: I grant that. It's these more diffuse, abiological systems . . . the sun is a hard step to take.

RUPERT: The sun has a very complex resonant pattern of magnetic fields with cellular vortices throughout its whole surface. It's a complex system of probabilistic turbulences and resonances with complete polar reversals about every eleven years, at the time of sunspot maxima. There's a physical interface—if a mind has to have a physical interface—that is an electromagnetic one at that. The Solar System as a whole involves all the planets, all the gravitational interactions, and the electromagnetic field of the sun, in which everything is made manifest through light. This field includes us sitting here in this room and everything else that's illuminated by it.

If light and vision are associated, as we started this trialogue by considering, then all things illuminated by the sun may in some sense be seen by it. The sun is in many cultures called an eye. In Malay, for example, the word for sun is *mata hari*, "the eye of the day." On the great seal of the United States, shown on every dollar bill, there is the Egyptian symbol of the Eye of Horus—the radiant eye, the sun—both a seeing eye and an emitter of light.

This brings us back to our starting point, the relationship between light and vision. The Eye of Horus is one symbol of their intimate relationship, and it's no coincidence that it is one of the principal amulets still used in Greece and other Mediterranean countries to ward off the "evil eye."

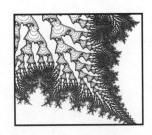

*In all times and all places, with the possible exception of Western
Europe for the past two hundred years, a social commerce between
human beings and various types of discarnate entities, or nonhuman intelligences, was taken for granted.*
   —Terence McKenna

*Entities exist in many realms. There are the realms of the dead, the
realms of dreams, and the realms of the imagination. There are also
the spirits of animals, the spirits of the Earth and solar system and
stars, and the angelic stellar intelligences. There are spirits of each
species of plant or mushroom, each with its own way of being,
its own way of seeing and experiencing the world, of participating
in the whole.*
   —Rupert Sheldrake

*Traveling up the great chain of being toward the world soul, we may
get in touch with things that precede any capability of verbalization,
that seem to reach out for contact, that are learning to communicate
in a language we can understand.*
   —Ralph Abraham

6

ENTITIES

TERENCE: This discussion will revolve around the exotic theme
of discarnate intelligences and nonhuman entities. These entities seem
to occupy a kind of undefined ontological limbo. Whatever their status
in the world, their persistence in human experience and folklore is
striking. The phenomenon of their existence is not something unusual
or statistically rare. In all times and all places, with the possible exception
of Western Europe for the past two hundred years, a social commerce
between human beings and various types of disincarnate entities, or
nonhuman intelligences, was taken for granted. This could have been
as simple as the Celtic farmer's wife leaving out a pitcher of milk for
the faery folk, or it could have taken more elaborate forms.

A second aspect of this theme is the tremendous variety of these
entities. We're talking about a kind of parallel taxonomy in another continuum in which there are djinns, afretes, water nixies, boulder grinders,
gnomes—and this is only the list within the context of the European
imagination. Once we add in the viewpoints of various cultures on the
potential for nonhuman life forms, we have a truly vast array of peculiar
creatures, all expressive of a very fundamental belief system that seems
to be inherent in the human condition.

Before we get into the history of this idea, it might be good to simply
review the logical options that are open to us in examining phenomena
of this sort. There are basically three. The first option is that these
entities are rare, but physical, and that they have identities somewhere
between the coelacanths and Bigfoot. They potentially could be imagined
moving from the realm of mythology into the realm of established
zoological fact, and this has in fact happened in some rather unspectac-
ular cases. This is by far the least interesting position. For example, the
yeti is a creature that refuses to declare whether it is simply a rare member
of the ordinary taxa of this planet or something quite different.

The second option that lies before us when we look at the ontological status of these entities is what I would think of as the Jungian position. To demonstrate it, I'll simply quote Jung on the subject of sprites and elementals. He calls them "autonomous fragments of psychic energy that have temporarily escaped from the controlling power of the ego." This is what I would call the mentalist reductionist approach to discarnate entities and intelligences. It says that they are somehow part and parcel of our own minds, their existence dependent upon our conceiving them as objects in our imagination, however pathologically expressed.

RUPERT: In other words, this is the humanist position that all gods, entities, and so on are simply projections of our own minds.

TERENCE: Exactly. The mental projection theory. "Escaped from the controlling power of the ego"—it's a wonderful image.

The third and obviously most interesting possibility, but the one fraught with argumentative pitfalls, is that these entities are (1) nonphysical and (2) autonomous in their existence in some sense. In other words, they actually carry on an existence independent of their being perceived by human beings. This is the classical position of those who have had the largest amount of experience dealing with these entities: the shamans, ecstatics, and so-called sensitive types.

This position poses a tremendous barrier for the scientific and Western mind. The eradication of spirit from the visible world has been a project prosecuted with great zeal throughout the rise of modern science. An admission that this project overlooked something as fundamental as a communicating intelligent agency co-present with us on this planet would be more than a dangerous admission of the failure of an intellectual method. It would pretty much seal the bankruptcy of that method.

Science has handled this problem by creating a tiny broomcloset within its vast mansion of concerns called "schizophrenia," deeming it a matter for psychologists, not the most honored members of the legions of the house of science. They have been told to "take care of this problem, please." This is where we get the Jungian mentalist reductionist model. What's interesting about this model, which is the reigning model concerning these entities, is that its appeal is in direct proportion to a person's lack of direct experience with the phenomenon that it seeks to explain. In other words, anyone who has ever encountered a discarnate intelligence knows that this is a woefully inadequate description of the phenomenon.

Before I close, I want to make one digression to drive home the point that this phenomenon is not simply the pursuit of dilettantes or obscurantists. If we examine the history of early modern science, we discover that some of the major movers and shakers were being guided and directed in the formulation of early science by discarnate entities. John Dee, the great flower of Elizabethan science, actually had commerce with angels and all sorts of entities of this type over decades. No less a founder of modern scientific rationalism than René Descartes was set on the path toward the ideals of modern science by an angel who appeared to him in a dream and told him that the conquest of nature was to be achieved through measure and number. This enunciation, which is really the battle cry of modern science, first passed through the lips of an angel!

There is also the well-known example of Kekulé, the discoverer of the benzene ring, who dreamed of the uroboric symbol—the snake taking its tail in its mouth, the ancient symbol of eternity—and understood that it was the solution to a molecular structure problem that he'd been searching for. This aspect of science, the fact that much of its premises have been transferred to mankind from the hidden realm of higher intelligence, is completely suppressed in its own official story. The official history tells the story of rational thought, of conquering the dark world of superstition.

I think as we look at these entities and try to place them in the context of human society in order to understand what they can do for us, we need to look toward the shamanic model. Here the spirits are not only identified but are seen as "helping," and a symbiosis is envisioned between ourselves and an invisible world with higher intent. This is what has been lacking in the expression of modern science as a social force. It would be a good idea to inculcate within any future model of society whatever wisdom and insight is represented by these forces.

It wouldn't be fair to close without mentioning that science's aversion to the irrational is something it inherited from Christianity. All of the voices of nature, of the sky and the Earth, were suppressed by Christianity in favor of the mystery of the Trinity. I interpret what's going on at the present moment as a rebirth, or a rise in volume, of the voices of the elementals. This seems to me part and parcel of the ecological crises of the planet. The planet is attempting to speak. Everything that can signify is reaching out toward humanity to try to reclaim us for the family of nature from the rather pathological trip we've been on for a long time. The elementals, the voices, the promptings of discarnate entities are to be carefully considered and studied. We are wandering in a wilderness, and they are a prompting voice.

RALPH: I'd like to focus on what you said about the suppression of the irrational by science and the trinity of Christianity, which came directly from the Neoplatonic tradition. I don't mean the Trinity of Father, Son, and Holy Ghost, but the trinity of body, soul, and spirit. The body is essentially Gaia of the Orphic trinity; it is ordinary reality, the physical universe of matter and energy. The soul is the oversoul or world soul, the parent of the individual souls that inhabit us. The spirit is a kind of elastic medium between body and soul, like the logos or the morpho-genetic field. Perhaps discarnate entities are evolving structures in the spirit, which contains information and intelligence. We can identify this as the top layer. There may be a hierarchy of these forms, which become increasingly anthropomorphic as they descend from the soul level to the body level. The spirit interpolates between the infinite complexity of the world soul, which includes our minds, and the relative simplicity of the world body, which includes our brains.

An abstract, cognizant entity may be impossible to know directly because of the complexity of the multidimensional spheres of the world soul. But when the entity structure descends, diffusing down through spirit, it becomes increasingly simple and develops more and more into cognitive forms that belong to the human mind in its evolutionary resonance with morphogenetic fields. At these lower levels, the entity is forced into representations that are culturally dependent such as faeries, dakinis, elementals, and so on.

This is a Sheldrakian interpretation of the Neoplatonic trinity of Plotinus. It has plenty of room for the plethora of entities within the spirit, just as Rupert makes plenty of room for biological species in the morphogenetic fields that guide biological morphogenesis. This model allows for a spectrum of different forms of what are essentially the same entities in different representations. Entities might be timeless in the celestial sphere, but they descend into a lowest representation as a cognitive map in our own consciousness, a map that depends on our evolving paradigm, our worldview. For this reason, there are different cultural flavors: elves, faeries, the pantheon of gods, and so on.

Early Christianity had this model. The Bible and Apocrypha are full of angels, archangels, and other entities. I think the Neoplatonic model is a context into which we could place all of these different representations. We could include the saints of all the world religions, the local deities, and all the multifarious representations of entities. The spectrum of representations corresponding to a single entity is essentially what Jung called an archetype.

TERENCE: Another set of questions is raised if you believe that these entities are nonphysical and autonomous. Are they related to us beyond the fact that we can share the same communication space? Are they somehow related to our own existence, not in the sense of being dependent upon it, but in some other way? All this talk of soul and spirit leads to the question of the relationship of the dead to these discarnate entities.

Did you know that the dogma of purgatory in Christian theology was not created by theologians in Rome but by Saint Patrick in an effort to make Christian doctrine more commensurate with Celtic folk beliefs? The faery faith that was in place when Patrick landed in Ireland held that dead souls coexist with us invisibly in ordinary space and can be seen by people who have a special ability. Saint Patrick turned that notion into purgatory, and he was so successful in the conversion of Ireland that theologians in Rome later wrote this concept into general church dogma.

RALPH: The Celtic people came from Eastern Europe, where they had contact with the cradle of civilization, as it's called. The view of the Underworld introduced into church canon by Saint Patrick already existed in Babylonian, Sumerian, and Ugaritic models. In the Sumerian myth of Inanna and Dumuzi, for example, Inanna goes to speak with her older sister, Ereshkigal, the queen of the Underworld. This story of a journey to the land of the dead is very old and runs very deep. The Underworld, as an aspect of the mythological side of birth and death, is part of one of the deepest layers of the mythic tradition. In a cave in Iraq, the dead were buried with flowers for the afterlife before 40,000 B.C.

The Neoplatonic model of spirit filling the void between body and soul may have a prehistory in which spirit was visualized as light. Thus, representations of entities have occurred in metaphors of light, particularized in whatever forms were available in a given culture. I believe the myth of the Underworld as purgatory, which goes back at least four thousand years to Sumer, is a displacement from the void between us and the sky—from the middle realm of spirit between the terrestrial and celestial spheres. Perhaps this has to do with the patriarchal takeover and the creation of the unconscious, or perhaps it has to do with our perpetual difficulty in dealing with death.

TERENCE: It occurs to me that all of these discarnate entities would be but dancing hallucinations before us if not for their ability to address us in languages that we can understand. Because they speak, we instantly transfer to them a whole new power and importance. They are transferring information from somewhere to us, only a very small percentage of which we are able to do anything with.

A traditional notion as it relates to elves and gnomes is that they are artificers of some sort—master artisans working with metals and jewels. Shamanism begins with a kind of deep penetration into early metallurgy. In this process, the smith and the shaman are twin brothers linked together in the extraction of energy from matter. This "whispering from the demon artificers," as Jung put it, has led us into technological self-expression and perhaps into self-expression per se.

The fact that these entities speak to us and we understand them is very puzzling for the rationalist. Most modern thinkers label this dialogue "schizophrenia" and put it away in a small isolated room and leave it there. However, a fair reading of the history of modern science, as Ralph earlier pointed out, shows that the irrational, in an objectified form, is very active in the process that we call history. We don't like to admit it, because we're committed to an official philosophy of reason and casuistry.

RALPH: There are two different theories going on here in terms of information transmission. In the horizontal theory, culture diffuses. For example, the agricultural revolution traveled from Anatolia in 7000 B.C. to France in 4500 B.C. to Britain in 3500 B.C. and so on. In the vertical theory, future evolution and past history depend on inspiration. There is a need to make shamanistic journeys, to travel vertically up into the logos, the world of spirit, where information is stored and is in a process of evolution. This information might be perceived as teachings from elementals, as in the teachings about metalworking received from faery folk.

In this vertical theory, we would expect the faery folk to show people how to work bronze, for example, in many different places at the same time. The Chalcolithic revolution could have been a simultaneous metamorphosis throughout the inhabited world if enough people had done vertical traveling. People like Descartes and Kekulé and John Dee were open to this vertical dreamworld, drawing information down from the celestial spheres in order to reinspire their own generations with this sacred knowledge.

These two models are compatible, and we may combine them into a single map that portrays shamanic journeys vertically and cultural diffusion horizontally. This map has two dimensions, whereas modern science and its paradigm have only allowed for horizontal diffusion.

TERENCE: Why do you suppose modern science became so adverse to these phenomena at the same time that there was such a zeal for the cataloging and description of all the productions of ordinary nature?

RALPH: I don't know. There was a fall. It happened in the century between John Dee and Isaac Newton in England. We can study everything that went on and still not understand much of what happened in this short period of time. None of the developments in the scientific enlightenment seemed to be explicitly adverse to angels. Newton believed in astrological alchemy, which embraced the significance of the stars, the hand of God, and the reality of the trinity of body, soul, and spirit. Descartes and Newton, whom we blame for our mechanistic paradigm, were full of the spirit. Of course, they dared not speak openly of these things because of the terrorist repression going on around them in the form of inquisitions, religious persecutions, and riots. Giordano Bruno was burned at the stake in front of an enthusiastic audience of three hundred thousand people in Rome on Easter Sunday because he refused to accept that the world was finite. Galileo was forced to recant for similar reasons.

Somewhere in this story is the basis for the rejection of the spirit by science. Why the church rejected it is another question.

TERENCE: The whole ambience of the world of Renaissance magic was one of lining up resonant incenses, minerals, and colors to call down stellar demons. Out of the Renaissance came modern science after the pact was made.

RUPERT: Starting with the astronomical revolution.

TERENCE: Yes.

RUPERT: I think discarnate entities are principally experienced by people in their dreams. In dreams, we travel in strange realms, we meet people who are dead or from different parts of the world, we enter strange situations and have unpredictable experiences. Our dreams exist in a kind of autonomous realm. The reductionist theory is that this realm is part of our own psyche. The more traditional theory is that we travel out of our bodies and enter what theosophists have called the astral plane. Some people have dreams of angelic beings. Others have nightmares. Heaven's pageantry covers many different regions. People throughout the world believe these regions are autonomous and that when we dream we travel out of our bodies into another realm.

Entities exist in many realms. There are the realms of the dead, the realms of dreams, and the realms of the imagination. There are also the spirits of animals, the spirits of the Earth and solar system and stars, and the angelic stellar intelligences. There are spirits of each species of plant or mushroom, each with its own way of being, its own way of seeing and experiencing the world, of participating in the whole.

All of these things are part of the shamanic fauna: the wolf spirits, crow spirits, other animal spirits, plant spirits, nature spirits, water spirits, mountain spirits, tree spirits, and so on. If you become like a hawk, you fly like a hawk, you see like a hawk, and you take on a hawk-like quality of being. These sort of spirits are biologically grounded. The angelic spirits are rooted in actual stars and planetary systems and galaxies. The whole realm is a system of intelligences that in some sense or another have a bodily aspect or were at one time in bodies, like the departed.

Are all entities grounded in bodily aspects? Or is there a free-floating, totally separate realm of entities of an entirely autonomous nature?

TERENCE: Perhaps everything at one point passes through matter, and that passage allows its eternal existence in an animate but discarnate realm of pure form. In this sense, biological existence would be fraught with the intimation of immortality.

RALPH: I would rather speak of animal souls than animal spirits, preserving the concept of soul for the ultimate end of the great chain of being. Spirit is a sort of elastic medium connecting it all. Spirit may be the venue of our travels in dreams and shamanistic journeys because we are unable to reach, in consciousness, all the way to soul. On the soul level, everything is connected up and all is one, as in the oversoul of Emerson and Thoreau. This great pancake in the sky participates in the material world by ripping off a piece of itself to incarnate in matter. In this view, which is the essence of the Hermetic tradition, everything has soul and souls are permanent. Their occupation as animals or rocks or trees is temporary.

In this Hermetic view, we may have the best chance to understand ourselves and our history. History on the scale of the world soul is a process of morphogenesis. Incarnation is the materialization of the morphic form, the entity, in the body. It is the morphic resonance of soul and body. Spirit is the abode of the entities, which are particulate aspects of morphic forms. The interaction between these different planes has been described as a resonant wave phenomenon.

When the whole biography of Gaia is seen in this Hermetic view, it may be possible to get back in touch with soul and be guided into the future. Traveling up the great chain of being toward the world soul, we may get in touch with things that precede any capability of verbalization, that seem to reach out for contact, that are learning to communicate in a language we can understand. The corn circles in England, for example, apparently are a kind of semiotic communication in which the cornfields, as organisms in the Gaian soul, are trying to speak to us. They

Figure 9. A Crop Circle. *An example of a "crop circle" formed by the flattening of the crop. The sides of the triangle are about 180 feet long. This spectacular formation appeared during the night of July 16, 1991, in a wheat field near the ancient fort of Barbury Castle in Wiltshire, England. A few crop circles are known to be hoaxes, but in England in recent years there have been more than two thousand formations of unknown origin. Reproduced courtesy of Richard Wintle. Calyx Photo Services © 1991.*

are developing an alphabet, little by little, just as we developed cuneiform script.

TERENCE: If there were no wind . . . wouldn't that be wonderful, if you were standing there watching and you could feel that the air was still and yet you saw the corn plants lie down one by one? Then you would understand who's behind this phenomenon. It's the Earth.

RUPERT: If it turns out that the corn circles are in fact what the most persuasive school of rationalistic explainers believes they are—namely, focused whirlwinds—they may still be the world soul communicating with us. Suddenly, whirlwinds take on entirely new patterns of focus and write new patterns. There's a spirit behind the whirlwinds that really expresses something, as many traditions have always believed.

RALPH: Gaia uses whatever means of writing she can find. If wind is necessary, if electromagnetic fields are necessary, then so be it.

TERENCE: The spiral form of the whirlwind is probably an extremely complex organized entity, an expression of the ordered morphology of the galaxy, resembling DNA. It is perhaps a higher form of life, but still manifesting the spiral energy form. Just as early scientists were shaped in their ideas by conversations with entities, perhaps we are on the verge of some kind of communications breakthrough. Perhaps, in order to have commerce and treaties and the exchange of patents with the invisible entities behind the corn circles, we need to bring to them all we've learned. Perhaps, applying differential calculus and the theory of hyperfine reactions, we may be able to work out a different treaty than we could have when we only understood blowguns and canoe manufacture.

RUPERT: That's where mathematical modeling comes in.

RALPH: I have in mind a mathematical model for the world soul, the spirit, and the mundane body that I would like to run as a video game in arcades or at Disneyland. The bad habits of science in the past four hundred years have had the unfortunate effect of depriving us of such a mathematical model. Our understanding of the material world, Gaia, or of the universe, the body of it all, is more advanced than our understanding of spirit, souls, faeries, or angels.

TERENCE: I wonder how accurate and reliable are the maps that occultists have accumulated in the past couple of hundred years. What struck me, when I first read A.E. Waite's *Ceremonial Magic*, was the classification of entities as lieutenants, generals, and majors. Each was assigned different metals, each its own sigil, and each offered its own dedicated incenses.

RALPH: A.E. Waite is an excellent example of the Hermetic view of

spirit/logos/m-field as an elastic medium between body and the end of
the great chain of being, soul. His model was inherited from the revelation to John Dee of an earlier magic, originally from the literature of
pre-Christian Jews and Chaldeans in Jerusalem and Alexandria. According to the Merkabah mystics, there are eight heavens that can be
visualized as concentric spheres. In the eighth heaven is God in his/her
own castle, and along the way there is, in odd places, a chariot with
wheels. Holding up the chariot is a being with four faces, one facing in
each direction. To enter the chariot there are four gates, each with eight
guardians, four on each side, and each with passwords and special signs.

RUPERT: What personal experience do we have of entities? Have you
ever met one? Could we try and meet one by carrying out an appropriate
ceremony or invocation in a suitably receptive pharmacological or
meditative state? If there are star spirits, as I believe is perfectly possible,
it's very likely that we can invoke them by various kinds of magic. What
kind of information would such beings impart? We know that ancient
civilizations had widespread beliefs about the particular properties of
various stars. Most accounts of angels describe them as innumerable as
the stars. The connection between angels and stars is very explicit in
the Christian tradition.

TERENCE: In a suitably receptive pharmacological state, the stage
suddenly becomes crowded with stellar demons, Earth angels, and what
have you.

RUPERT: Do we know for sure these are stellar demons? Have you
ever connected your experience of a demon with a direct look at his
resident star?

RALPH: We have this experience in a limited way. The sky is sort of like
a Rorschach test. There are a lot of dots that can be connected in many
ways. When there is a tradition of connecting them in a certain way, there
is an astrological tradition, a star mythology, an asterism. Pictographs,
petroglyphs, and cave paintings include drawings of asterisms. The Greek
myths are projected onto the celestial sphere in the asterisms we call
constellations. The word *myth* is from the word *mythos*, meaning "lyric"
or the words of songs from rituals. Myths gained the power they now

have in our conscious and unconscious lives through their secondary role in rituals. The old rituals were actually means to bring the celestial forms down from the soul level, through the spirit level, and into the body. This is where star magic worked successfully to empower our evolution, and this is as sure as we can be of the existence of star demons.

TERENCE: In our culture, we tend to move into cities that push nature away from us. In our mental environment, we do the same thing. Most people live within a very conventionalized set of notions that are deeply imbedded in a larger set of notions. When we go to the physical edges, such as desert, jungle, and remote and wild nature, and when we go to the mental edges with meditation, dreams, and psychedelics, we discover an extremely rich flora and fauna in the imagination. This realm is ignored because of our tendency to see in words, to build in words, and to turn our backs on the raging ocean of phenomena that would otherwise entirely overwhelm our metaphors.

RUPERT: If we ask what has caused this blindness, we might answer that it's the satanic spirit of science. In the seventeenth century, the spirit of Satan was portrayed in Milton's *Paradise Lost*, with a whole taxonomy of various demons and fallen angels that acted as malevolent powers, such as Mammon, the demon of commercial greed. The primary sin of Satan and of the other fallen angels like Mammon was pride, the turning away from God toward their own self-sufficiency. This was the beginning of the whole humanist illusion that turned away from the spirit world and declared humans to be self-sufficient. From this point of view, all gods, demons, and spirits are projections of the human mind, creating a kind of anthropocentric universe.

TERENCE: Humans are said to be the measure of all things.

RUPERT: This is humanism. To adopt the alternative tradition of animism and to recognize the living spirits and souls of all nature is profoundly repugnant to humanism, yet it is the common ground of all human civilization, thought, and tradition. As in Goethe's *Faust*, the paradigmatic scientist sells his soul to the devil in return for unlimited knowledge and power. The guiding spirit of modern science, according to the Faust myth, is a satanic demon, a fallen angel called Mephistopheles.

How seriously do we need to take the idea that our whole society and civilization is under the possession of such a spirit, worshiped through money and power? Milton describes Mammon in *Paradise Lost*:

> Even in heaven his looks and thoughts
> Were always downward bent, admiring more
> The riches of Heaven's pavement, trodden gold,
> Than aught divine or holy else enjoyed
> In vision beatific: by him first
> Men also, and by his suggestion taught,
> Ransacked the centre, and with impious hands
> Rifled the bowels of their Mother Earth
> For treasures better hid.

(1.680-688)

This is an accurate description of our whole civilization. How much are fallen angels actually guiding and perverting the progress of science and technology? Is a great war between the good and evil angels being acted out on Earth? We hardly know how to think or talk about such possibilities since they are so alien to the official, standard models of Western history.

TERENCE: Returning to the subject of discarnate entities, I keep going back to this thing about language. It's as though the field of language itself must be prepared for communication with these beings. In the West, there has been a peculiar stiffening of language against the ability to express this kind of communication, but it is beginning now to break down.

RALPH: It's true. We have to misuse our language even to talk about these things.

TERENCE: Linearity in print and thought has made language unable to deal with the invisible world in any meaningful way, except as pathology. Now this invisible world is returning to the language through people like us with one foot in each world.

The human mind is haunted both by the many presences sensed within the self and by a confused sense of self. Wherever we turn in the world of nature and the psyche, we encounter life, animation, and a willingness to communicate that confounds the fragile pyramid of boundary consciousness and human values that have emerged over historical time through the suppression of our intuitions.

I've taken the position that these entities we encounter are nonphysical and somehow autonomous. Ralph, as I understand him, accepts this view but anchors it into the Neoplatonic trinity of body, soul, and spirit. From this point of view, these entities are inhabitants of the spiritual domain of the logos. They are the logos become self-reflecting and articulate. Rupert correctly points out that it's in the realm of dreams that we most commonly encounter entities, and he further suggests that behind these entities is the controlling agency of the world soul. His notion is that the world soul actually communicates to human beings through the production of forms that we interpret as the denizens of an otherwise invisible and mythological world.

Our collective conclusion seems to be that nature, both in whole and in many parts, is magically self-reflecting and aware. Encountered in its most rarified expression, the world speaks to us, and we, as scientific rationalists, are confounded. Nevertheless, it is for us to mold our models and theories to the world as it presents itself in immediate experience, not as we would have it in some grand and sterile abstraction. The elves and gnomes are there to remind us that, in the matter of understanding the self, we have yet to leave the playpen in the nursery of ontology.

*By giving people a less restricted choice of addictions, we can
cause some growth in consciousness and some shrinkage of the
unconscious. I think it's not necessary to make the whole unconscious become conscious. If we can't undo this bifurcation in which
the mental curtain developed, we can at least rearrange the furniture a little bit.*
   —Ralph Abraham

*It seems that the focus of attention or awareness is quite narrow. . . .
It is as if sensory awareness takes place on the surface of a largely
unconscious system. On the shimmering surface we notice differences, but below the surface is habituation and unconscious
habit. This is at the heart of all living systems.*
   —Rupert Sheldrake

*The reclamation of the unconscious has to do with directing
attention toward understanding time. Time is apparently the body
of the unconscious.*
   —Terence McKenna

7

THE UNCONSCIOUS

RALPH: I would like to start with three great bifurcations in the history of consciousness. The first is hypothetical: the unconscious was created one afternoon by a bifurcation in the collective mind. At this point, a curtain was created that divided the conscious from the unconscious forever, like an Iron Curtain in the species mind. The second great bifurcation in the history of consciousness was the Fall from the Garden of Eden, which was the origin of evil. The third great bifurcation was the defeat of the dragons of chaos by the gods of law and order.

With respect to the first bifurcation, we could speculate that in the prehistoric period the conscious mind developed from the unconscious mind. Or, it may have been the other way around—that both existed always and there was never any division.

However, for the sake of discussion, let's say that consciousness existed always and that the unconscious was created by a recent bifurcation, a prehistoric consequence of a natural law of evolution. Because of peculiarities in the evolution of the conscious mind, more and more perceptions were regarded as illegal in the context of culture and civilization and the current worldview. In time, as more things from above the curtain were forced below it, the unconscious continued to grow as the conscious mind correspondingly shrank. In this continuing evolution and growth, there were further bifurcations, and all of the fundamental deck furniture of the unconscious, including dreams and archetypal symbols, emerged.

One virtue of this theory is that it allows us to see our individual conscious minds evolve from a universal conscious mind. We may then speak of the consciousness of animals, of plants, of Mother Earth and Father Sky, and of the whole nuclear family of all and everything.

The second bifurcation dealt with the creation of evil, the expulsion of Adam and Eve from the Garden of Eden, the fall of Lucifer, and the empowerment of the negative. Lots of dichotomies emerged from

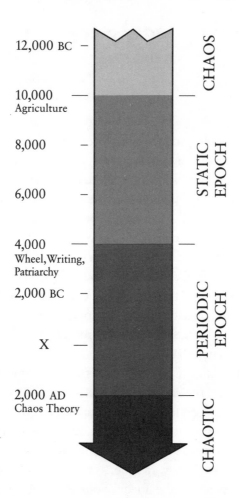

Figure 10. The Three Great Bifurcations of Cultural History.

unity by a process of mitosis, or bifurcations, in the world of ideas.
We could try to relate these events to each other in time, or in space,
or as waves, or as a holistic space/time pattern.

Number three was the bifurcation of chaos and order in the cosmos.
As mentioned previously, this horror story was ritually enacted during
the Babylonian New Year festival. The king played the role of Marduk,
who divided Tiamat, the dragon goddess of chaos, in two with a sword.
She opened like an oyster shell, forming the sky and the Earth.

These three bifurcations are related. When the myth of Marduk became dominant, a negative charge was put on the serpent image. The snake is an essential symbol of Mother Earth herself. The consciousness of Gaia came to be associated with Hades, the underground, death, evil, Pluto, and so on. The repression of chaos reinforced the bifurcation in consciousness that gave us the unconscious, which has gained ever since. Thus, the human species has been heading more and more into the evil, the demonic, and the evolutionarily unsuccessful forms of society.

This is the chaos theory of the bifurcation of mind or the creation of the unconscious.

TERENCE: What do you think caused this bifurcation?

RALPH: I don't know. Historically, there is this horrific story, the Babylonian creation myth—a myth as lyric opera and ritual. This myth had a Sumerian precedent. Towns in the early stages of the urban revolution coalesced around different gods and goddesses. The form of a goddess or god was materialized in the form of a city. This was the origin of the sacred city. There was therefore a morphogenesis in the full complex of population movement, urban technology, city design, religion, ritual, and so on. The successful god or goddess concepts were attractors that created cities. These cities resonated with other cities, and through this syncretism empires were created.

There are books about the serpent and dragon images in mythology. Many cultures had dragon images, and they were all associated with chaos and the Goddess. Then, suddenly, there was Saint George killing the dragon and Saint Michael with his foot on the dragon's throat. All this energy was given to killing dragons in Babylon, in Canaan, in Celtic mythology, in Christianity, and in science.

I think it's possible that this particular bifurcation had to do with the patrilineal necessity of knowing the identity of a child's father and with the rejection of the Dionysian ritual in which sexual license was an important part. The patrilineal structure involved the creation of the nuclear family, which brought about the rejection of the Goddess, the dragon, and chaos. Now, I think it may be important to undo this repression and have Saint George and the dragon get it on together in a May Day celebration where Dionysian elements are accepted.

RUPERT: So you're in favor of a kind of "fall" theory of the unconscious.

RALPH: I'm proposing it for discussion. I'm trying it out. It's an hypothesis which, if we accept, we could seek also to escape. Somehow, we'd have a mechanism to escape because we'd have a historical map of the creation itself.

TERENCE: How would this escape happen?

RALPH: We could try to bring unconsciousness back into consciousness. We could take the deck furniture from the lower floors of the boat back up to the main deck where it belongs. In chaos theory, which is the current paradigm shift, science, which is now the main temple of law and order, has to eat humble pie and accept chaos again. This is apparently happening more or less accidentally. Suppose we were having this talk in 1960, and we decided we would try to do something intentional that would restore Tiamat to her throne. We'd hunt around with the computer people, we'd locate a closet chaos pioneer like Ueda or Lorenz, and we'd popularize his model. We would purposefully revolutionize the sciences in 1963 instead of waiting for this to happen spontaneously in 1973.

I'm not recommending an intentional intervention. I'm saying that bringing some of these things into consciousness would have a big effect on future generations. This discussion alone could be all it takes, you see, to bring something back from the basement and restore it to consciousness, starting a new upward spiral somewhere in the evolution of spirit.

TERENCE: The suppression of psychedelics occurred at the same phase because orgies driven by the psychedelic style of religious celebration completely frustrated the desire to identify male paternity.

RALPH: The rejection of psychedelics and shamanism could be likewise associated with these three bifurcation events. Only so-called primitive societies have shamans and perform psychedelic rituals on a regular basis. Their survival in fringe societies is a vestigial tail. They are survivors of the fall, of the loss of a whole complex of ideas.

TERENCE: The shift from the psychology of a psilocybin-mushroom cult to the psychology of an intoxicating-mead cult, for example, would have led precisely to the difference in interpersonal dynamics that would have shifted a group from partnership relationships to patriarchal relationships. I think that, over time, the mushroom-using people in the archaic Sahara were transformed unwittingly into a mead cult by a scarcity of mushrooms and by the simultaneous recognition that honey was the potential preserving medium for the mushrooms. The effort to preserve mushrooms in honey and to concentrate the use of them into occasional large festivals eventually evolved into the mead festivals, where mushrooms became no more than a flavoring in mead and ultimately became no more than a memory.

RALPH: First the mead, then the patriarchy?

TERENCE: The patriarchy and the mead evolved together, along with the concern for male paternity and the suppression of orgies. The critical decision was the choice between fun, on the one hand, and the full knowledge of the flow of one's genes, on the other. Once male paternity became an important issue, then the concept of "me and mine" came into existence: "my women," "my children," "my food," "my weapons," "my land." This is the attitude that the orgiastic, psychedelic, boundary-dissolving mushroom religion held at bay.

RALPH: This is an alternative to the theory that patriarchy came about because of a wave of invading Kurgens.

TERENCE: It would have happened quite naturally, beginning with a flourishing mushroom cult on the plains of Africa and the gradual climactic drying that created the seasonality of the mushroom. There were wonderful huge seasonal celebrations, and the psychology of the group remained basically intact. Then, as the drying continued and the waterholes became further and further apart, the need to preserve the sacrament and to spread it ever more thinly became strong. Desertification of the Sahara also occurred through overgrazing by patriarchal, ownership-oriented pastoralists who worshiped the male thunderbolt god. The actual use of the land, in the eradication of partnership-oriented hunter-gatherers, created a desiccated environment unable to produce mushrooms. Thus, the psychedelic mysteries slowly faded from

Figure 11. A Mushroom Artifact. *Such an object is assumed to indicate a Meso-american cult of mushroom use that flourished more than two thousand years ago. From a private Canadian collection. Greif-Czolowsksi Photography © 1988.*

the occupied zone of the culture using them, and an only partially satisfying substitute took their place.

RALPH: Booze.

TERENCE: They went from an ecstatic Goddess cult of orgy to a drunken revelry of warriors and whores.

RALPH: It's a two-phase theory: the psychedelic partnership phase and the alcoholic dominator phase. These are drug-driven phase transformations repeating themselves in an endless flip-flop in the history of consciousness. There is historical evidence of this in Crete, where in historical times there was the debasement of Dionysian rituals into alcoholic revels.

TERENCE: Dionysius turned into Bacchus. The early Bacchus was a frightening figure. His practitioners fell into such frenzies of ecstasy that they were charged with devouring their own children. The late, hairy-footed, lascivious Bacchus is simply an image of alcoholic consciousness made concrete.

RALPH: Orpheus was a reformer who was trying to get the Dionysian religion back on the psychedelic track, away from booze. Such a reform happened again with Pythagoras, then with Buddha, and so on.

If we take this two-phase theory seriously for a moment, and we favor one phase more than the other, then we would be inclined to seek the mechanism of phase transition; an example of such a mechanism would be turning on the heat under water so that it will boil. This theory explains why alcohol is legal and marijuana is not. There is a self-preserving, phase-maintaining function within the alcohol phase.

TERENCE: It reinforces male dominance. If we analyze what alcohol does pharmacologically, we find that it actually diminishes sensitivity to social cueing. This is a technical way of saying that when you drink alcohol, you turn into a bore and a libidinally driven oaf. Your normal social judgment is impaired and you're prepared to make a sexual conquest. Most of the sexual neurosis of Western civilization can be traced to incidents of early sexual imprinting in the presence of alcohol. Alcohol is so spun into our simultaneous terror and attraction for the sexual experience that it has become an invisible part of our cultural legacy.

It's a matter of the way in which our relationships to the material world, especially drugs, promote or retard the expression of what we're calling the unconscious. One of the things that repels us all about drug abuse is that it is unexamined behavior. When we are addicted to heroin or television, we turn ninety percent of our lives into an indwelling in the unconscious.

RALPH: I feel that I'm unable to become totally unaddicted. I don't have to be addicted to alcohol. I can replace one addiction with something else. I can actually choose, reclaiming free will without rejecting addiction.

TERENCE: This is the kind of creature we are. The addiction to addictions that spawned this itch originally was no addiction at all but rather our natural connection into the hierarchy of Gaian information that was accessible to us when the nature religions were freely practiced. A sacrament is not a symbol. The day people were sold on the notion that a sacrament *was* a symbol, the umbilical connection to the logos was severed, and history was run off track.

RALPH: By giving people a less restricted choice of addictions, we can cause some growth in consciousness and some shrinkage of the unconscious. I think it's not necessary to make the whole unconscious become conscious. If we can't undo this bifurcation in which the mental curtain developed, we can at least rearrange the furniture a little bit. What else can be done, besides diminishing alcoholism, to advance consciousness, develop spiritual telescopes to look into the unconscious, and regain our wisdom?

RUPERT: I don't think the problem is that the unconscious is the result of a kind of conspiracy or fall but that nature works through the formation of habits. We know from our own experiences that habit formation involves habits becoming unconscious. We experience habituation, for example, when we go into a room with a funny smell. After a while, we stop noticing the smell. The same applies to background noise: we get used to it. We share with the entire animal kingdom, right down to the level of unicellular organisms, this tendency to become unaware of most of the environmental stimuli acting upon us.

It seems that the focus of attention or awareness is quite narrow. In general, what sensory systems do throughout the animal kingdom is sense differences—in smell, temperature, pressure, or texture. It is as if sensory awareness takes place on the surface of a largely unconscious system. On the shimmering surface we notice differences, but below the surface is habituation and unconscious habit. This is at the heart of all living systems.

I don't think there's a kind of sinister conspiracy of the unconscious. Perhaps a sharper awareness of the interface between the conscious and unconscious developed at one time—through mystical experience, psychedelic visions, and shamanic journeying—revealing realms of experience of which we are not normally aware. We can bring other realms of experience into consciousness through ritual practices such as the annual cycle of festivals found in all religions. Through ritual practices still found almost everywhere, even in the West—including Jewish festivals, the Catholic calendar of saints' days, and major festivals like Easter and Christmas—different dimensions are brought into awareness.

We can't be conscious of everything all the time. We have a limited focus of attention.

TERENCE: You're saying a calendar can be an engine for illuminating the unconscious?

RUPERT: Yes.

RALPH: I think the theory of habituation is a good model. In it, the mind is viewed as essentially unconscious, and consciousness is seen as a little window that can be directed at will—by ritual or voluntary choice—over a certain part of the unconscious, which would then become conscious for a day. The unconscious I spoke about before is a region of the whole realm of consciousness whose contents have become unavailable because they were declared illegal at a certain time in history. It's not possible to rove the window over this region any more.

RUPERT: Perhaps because it doesn't have its day. The orgy principle had its day, and it still does in the Hindu festival of Holi: people get high on cannabis, throw colored water at each other, have priapic processions, and overturn the social order. Similar principles are at work in the

Catholic world through Carnival or Mardi Gras. However, these festivals were suppressed in Protestant countries as a result of the Reformation. Chaos in the West no longer has its day, nor does the Great Mother have hers.

RALPH: By focusing our attention through special days and rituals, we can work at maintaining our consciousness much as we work at maintaining our garden. This is a good theory of the importance of holidays, but there is still something wrong. Every year Christmas is repeated, but people have completely forgotten its archetypal significance, even with the full regalia of angels on the trees and so on.

RUPERT: The archetypal delight of Christmas is experienced by young children waking up on Christmas Day and seeing stars and angels and presents, with glittering frost and fresh snow in the background. For them, the magic of Christmas works. It is, after all, a young children's festival because it celebrates the birth of the sacred child. This is the pole it touches in all of us: the sacred child. I think it still helps to maintain the basic pulse of the sacred year, however sluggishly, in a lot of people.

TERENCE: Notice that what you are suggesting is that the reclamation of the unconscious has to do with directing attention toward understanding time. Time is apparently the body of the unconscious.

RALPH: Time is the mediator of differences. Time is the enemy of habituation. Time is the strategy for subverting sensory inhibition.

TERENCE: Time is the theater of habituation as well. Time and attention used creatively can banish the unconscious.

RALPH: What about the dark part of the unconscious, over which it's illegal, difficult, perhaps even impossible to fly our plane to look out the window of consciousness? The experience of the window of consciousness is our heritage of perception, but it has been denied to us on the basis of religious authority and dogma maintained by ritual. We can't reclaim a dark part of our unconscious simply by a willingness to devote one day to attention, to variation, to observance of a difference, or to amplification of this part of the unconscious. It has just become unavailable.

Denial, I think, is a recent phenomenon, and it is a serious danger for evolution because, once an experience is denied, evolution is then shunted off its track. This kind of unconsciousness, I still think, was created at a certain time in history by a bifurcation. Somewhere around six thousand years ago, there was a special new mode of thought involving the prohibition of valid experience. This was the real fall, on the basis of which we now have real evil as a problem in life. The destruction of integrity and the death of nature are new problems with a history that has to do with denial.

After something has become not okay, it's very hard to reclaim it. For example, chaos is considered bad, and it's very hard to undo this damage.

TERENCE: What I hear you saying, Ralph, is that some fundamental boundary must be breached or dissolved in a dramatic way to confirm that we've entered a new phase of the human adventure. Perhaps the revolution in media and information technologies is the transparently transforming solvent that, by informing us concerning the real options available to human beings, can get us off the limited Western approach. It's important to create and make strong our mental community.

RALPH: I think it's possible to raise the frequency, to lighten the dark, through prayer. I'd like to see a resurrection of magic. Hopefully, many places that have fallen into disuse, where the garden is overrun with weeds, can be sweetened with banishment ceremonies, prayers, and so on. We need to learn this technology—to connect the star magic of the Stonehenges and astrology, for example, with the progress of daily life and political events.

We have great powers that aren't being used because we don't believe in them. The unavailable unconscious contains enormous power for doing good and can provide us with our only means for recovery.

The resacralization of space and time involves not only recognizing the sacredness of churches and cathedrals and traditional festivals, but recognizing the importance of sacred places everywhere and of every kind. . . . This involves a much more animistic version of Christianity and Judaism, a process I've come to think of as the greening of God.
      —Rupert Sheldrake

We need a unifying principle and attractor. The essence is an actual connection to the sacred. I don't think we can have an archaic revival by simply going backward. We have to carry our archaeology of knowledge to the point where we understand the essence of what took place in the past and then adjust it into modern forms.
      —Ralph Abraham

The purpose of science should be understanding, not only technique. We need to hold back while we assimilate what we know at our current level, not push relentlessly deeper into the application of techniques. That is rape. That is violation.
      —Terence McKenna

8

THE RESACRALIZATION
OF THE WORLD

RALPH: The story of my religious training and upbringing may
be appropriate to introduce our discussion on the resacralization of
the world.

My parents both came from Jewish backgrounds. My mother was
a regular temple visitor in her town until she married. My father had no
connection with Jewish traditions since his parents relocated from New
York to the backwoods of Vermont in their youth. There was never
mention of Jewish religious stuff in our house, as I think he developed
an aversion to organized religion of any sort. Occasionally, circumstance
would bring me into church while a service was going on, and I would
look around at the other people who bowed their heads when told to,
thinking, "I'm the only one who's not casting my eyes down; I'm the
only one who doesn't believe this stuff."

My father developed a kind of religion of his own, centered on the
family and on a concept of love that was fairly abstract. He used to say
that love was everything. Explicitly, he thought it was better to start
practicing religion when you were an adult, when you could acquire
whatever religious training you wanted, considering all options with an
open mind. I never considered all with an open mind, nor did I decide
whether to have one or none; I just didn't bother to think about it.

Practically everybody I knew either went to some kind of church or
had been brought up within one and had formally rejected it. From them
I got no evidence of any sacred knowledge outstanding in any of these
churches. Eventually, the sixties happened, and with LSD I had what
I would now identify as my first religious experience. Immediately,
I thought, as did many people in those days, that this was a religious
experience in the true sense of religion.

Reading history, I came to see that there had indeed been sacred
knowledge in the churches of Europe in the past. It was embodied in
architecture and music and had become extinct in the Western world

121

only recently. But the sixties were very preoccupied with politics, so these were just idle thoughts at the time.

When I got to India in the seventies, I found that the extinction of the sacred from organized religion hadn't occurred there. Instead, I found a living tradition that I recognized as sacred.

When I returned from India to California, about the time I met Terence, I began to think there might be a way to revitalize the Western tradition. Other people spoke about this idea: those concerned with yoga and with the adepts of the Middle Ages and Renaissance, Theosophists, and students of alchemy. In 1972, there was a distinct separation between political activity and concerns about the loss of the sacred from religion. In the course of time, it was suggested that a root cause of the political problems of the world was the loss of our connection to the sacred, within and without organized religion. When this connection was made, political activity had to address it.

People have different theories of evil and the causes of global problems. Because of the strength of my direct experience with the sacred, desacralization became and remains my favorite theory. I believe that the most important activity to save the world, or at least to move toward hope in that direction, is to recreate for some larger portion of humanity the lost thread of our connection to the sacred. This is the program that I call "the resacralization of the world."

As far as I know, there are only a few proposals for direct action in this direction. One is the resumption, in everyday society, of rituals that still exist for some native peoples—rituals such as tribal shamanism. Another is the revitalization of existing religion by attracting people back to the wonderful churches of the world. These churches have not only the architectural shape most appropriate for the communication of the sacred thread, but they have tradition as well, which is the morphic field of our connection with the sacred. We can reconnect the missing link through an archaeology of knowledge. We can attempt to find when and for what reason the light went out and then attempt to reforge the link.

A third proposal is for people to remain within the context of modern society but outside the organized churches, replacing churches with an equivalent institution based perhaps in the sacred arts of music, architecture, and painting.

A fourth idea, to which we all may have given too much attention, is the revitalization of science. This proposal acknowledges that science has replaced traditional Christian mythology with its own substitute mythology and that weaknesses in this mythology may be at the root of evil.

A fifth idea has to do with the reestablishment of the partnership society. The work of people like Riane Eisler suggests that the desacralization and devitalization of the church and organized religion is associated with the embodiment of patriarchal values within organized religion. This is a corollary of the mechanistic tendency of dominator societies.

I don't think everybody should take psychedelics or have a shamanistic experience in the Amazon jungle to rebuild their connection to the divine. However, churches and their rituals could be reinvigorated if a certain class of priesthood specialized in maintaining this kind of direct, Gnostic connection. The weakness of scientific mythology can then be fixed by somehow incorporating the scientific worldview into the expanded church view. There would be some sort of syncretism between the neopagan and the Judeo-Christian traditions, repairing the patriarchal weakness and restoring revelation with the inclusion of a new sacrament.

In short, I'm looking for something beyond the understanding of the virtues of the past. I'm looking for the basis of a program in the present, with the resacralization of art and music as an integral part of the program.

TERENCE: You're talking about an archaic revival: the resumption of ritual, the revitalization of existing religion, and feminism and the revitalization of partnership in there somewhere. Sometimes you seem to be talking about a revitalization of the forms of the fifteenth and sixteenth centuries, and sometimes about forms from much further back. All this business about music and churches is relatively recent.

RALPH: This loss has been going on for ten thousand years. What has been lost can't be revived in its original form. We may have to use video and film technology. The value of getting true partnership-society values into the church is that we wouldn't have to replace the church just because it's been on the wrong track for four thousand years.

Figure 12. A Green Man. *An example of a green man, with foliage coming out of his mouth. Such foliate heads are a common but mysterious image in gothic churches and cathedrals. This one is from the fifteenth century, in the church of Llangwm in Wales. Reproduced courtesy of Clive Hicks.*

RUPERT: Practically all of the Gothic cathedrals are still functioning: they still have sacred chants going on every day, and prayers are offered in a regular cycle, just as great temples have always done. These continuously active, sacred places have been sacred for centuries. The great cathedrals, like temples, are models of the cosmos. In the artwork of most Gothic cathedrals there are green men: mysterious vegetation gods that burst out everywhere with leaves coming out of their mouths. Many of the great cathedrals are dedicated to Our Lady, who in the Middle Ages was often seen as Wisdom, or Sophia. They have in their windows geometrical designs: threefold, fourfold, and fivefold mandalas. There are rose windows with extremely complex psychedelic designs and colors. By participating in the spirit of these cathedrals, we can reconnect with the sacred places of Europe.

Another important route for resacralization is through the practice of festivals and the sacralization of time. Some Jewish people I know are revitalizing the tradition of the Sabbath, starting on Friday evening with the lighting of the candles and the invocation of the Shekinah, the feminine presence of God in the home, in the world, and in embodied existence.

The sacralization of time is a way of reconnecting ourselves with religious traditions. I think the most important aspect of this process—because I agree with Terence about archaic revival—is to find the pre-Christian roots that underlie the timing and quality of existing forms and festivals. We need to ground the new religion in the old. In the present tradition, there's a continuous living strand that goes right back to the pre-Christian shamanic societies of Europe and the pre-Jewish shamanic societies of the Middle East.

This is a connection that can still work; it works for me. Obviously, the resacralization of space and time involves not only recognizing the sacredness of churches and cathedrals and traditional festivals, but recognizing the importance of sacred places everywhere and of every kind: holy wells and springs, sacred hills and groves, ancient sacred caves, and the archaic sacred places of the megalithic age. This involves a much more animistic version of Christianity and Judaism, a process I've come to think of as the greening of God. Theologians like Matthew Fox have exposed a hidden strand within the Judeo-Christian tradition that can authenticate and feed a new burgeoning of Christian and Jewish animism.

RALPH: But is it possible? I'm not sure.

TERENCE: Marshall McLuhan said it was inevitable. He felt that the shift from a literate phonetic-alphabet culture, which existed for us as recently as the 1940s, to the electronic culture would effect the ratio of the senses. The new ratio created by electronic media would be similar to that which existed in medieval Europe before the invention of printing. He based this notion of "electronic feudalism" on the idea that before the linear uniformity of mechanical print we had to *look* at each individual manuscript, because every manuscript reflected the hand of its author. After print was invented, we no longer *looked*; we *read*. Reading is a very generalized function in which we don't actually study each *e* and *l*.

With television and electronic media, we are again returned to the situation where we must *look*. We must assemble a gestalt, an image; we cannot simply read it. McLuhan felt that the consequences of this shift in the sense ratio would be global and immense—that it would cause the fragmentation and dissolution of the nation-state, which we certainly seem to be seeing in Eastern Europe and the Soviet Union. He said it would return people to a kind of homebound pietistic funda-mentalism—homebound because electronic media brings informa-tion everywhere. McLuhan predicted that all of these factors would conspire to create an era in which the Gothic model would be very strongly expressed.

From my own point of view, in my theory of the time wave, I see this coming in the mid-nineties and expect this period of time to have as many ambiguous aspects as did the Middle Ages. On the one hand, there was the glory of the Gothic cathedrals, and on the other hand, it was a time of wandering flagellants, pestilence, bigotry, suppression of women, hatred of outsiders, insularity, provincialism, barbarism, and so forth.

I think we're in an excellent position to experience something very much like the Gothic revival you're advocating. The important thing is to humanize the impulse.

RALPH: We need a unifying principle and attractor. The essence is an actual connection to the sacred. I don't think we can have an archaic revival by simply going backward. We have to carry our archaeology of knowledge to the point where we understand the essence of what took place in the past and then adjust it into modern forms. We have to acknowledge, for example, that the total population is larger now. That means recycling is mandatory. That means green consciousness, Gaia consciousness, will have to play a leading role in rituals performed on various days. The mandate is to create a new mythology that can organize different styles of churches on the path of convergent evolution.

RUPERT: I think religious reform cannot come from the traditional reli-gious hierarchy but must come through an attractive new movement that is practiced in people's lives and spreads through example.

RALPH: There's a real gap in the strategies for starting a new system between those based on archaic or pagan revival and those based on revolution in the churches or resacralization by religion. The traditional church denies the validity of archaic pagan forms, a denial carried to the extreme of revising history and pulverizing goddess figurines and statues. Besides partnership of the genders, local control, and a green politic within the ritual of the church, there must be acknowledgment of the essential religiosity of the pagan forms.

RUPERT: If there were to develop a true Mother Earth religion, it would probably have priestesses rather than priests, because its central figure is a goddess. It would relate human life to the Earth first and foremost. Our bodies return to the Earth. Such a religion would be into the whole material cycle, with little emphasis on the stars or the heavens. I think it would become very claustrophobic for a lot of people before very long. There would then be reborn a religion that brought in the aspiration toward the stars, the heavens, the greatness of the cosmos. We'd have a choice of religious images. We could choose, for example, between a god or a goddess of the heavens.

RALPH: We are envisioning a revolution of religion in which there would be priests, priestesses, local control, a renewal of meaning in rites and rituals, and so on. Unfortunately, there has been a continual decline in attendance at churches. The fact that a revival may be already underway is of no great use unless it becomes attractive. There has developed a pretty strong habit of revulsion toward churches and all their sins over these past centuries. Plus, there's the competition of scientism as a new mythology that is totally disjoint from the churches and attractive because of the power of its weapons.

RUPERT: One of the big emotional plusses of scientism for people is its sense of superiority. I converted to scientism at about age fourteen. A bit like you, Ralph, I looked around at everybody else praying or appearing to pray and imagined I was seeing from a higher point of view in which all this was superstition. Modern people feel superior to the part of religion that seems infantile and part of the past. However, a collapse of faith in scientism is happening all around us; a widespread public disillusionment with science has been developing.

RALPH: People are left stranded, having rejected the church in favor of science and having rejected science in favor of nothing. This is the dilemma of today.

TERENCE: Science and green politics can be sacralized through the psychedelic experience. The psychedelic person knows that the scientist dismissing people bowed in prayer is a poor fool. A green party that uses a mystical language, a psychedelic language, a language of integration with nature and emotion would have tremendous appeal. That's why Rupert is so keen on the *ayahuasca* cults of Brazil, because on one level they seek to preserve the rain forest and help out little people, but on another level they offer a psychedelic religion that makes claims on the imagination, the heart, and the world soul.

RALPH: How about celebrating the Eleusinian mysteries in the Cathedral of Saint John the Divine, or something of that scale?

TERENCE: It has to be understood that psychedelics are a way to the Gaian mind. They are not metaphors for sacraments, they are real sacraments, and their efficaciousness can have political consequences. A mystical political movement would become a crusade or a jihad. The energy of the attractor is so great in the mystical dimension that it creates a situation in which saving the Earth is not something we argue about, it's something we go out and do, like recovering Jerusalem for the pope.

RALPH: We had a big start in the sixties toward this kind of psychedelic/ green revolution. It failed for different reasons. For simplicity, I'll suppose that the failure of the hippie subculture was due to its repression.

TERENCE: It had no program, that's all. The only program was to end the war.

RALPH: But there were festivals, cults, and rituals that were changed to suit the locale, the sacrament, and sacred music. A lot of things we're talking about were actually happening in the sixties. But there was great and successful opposition to the program we're now envisioning. What about the desire of the current system to maintain itself by opposing the evolution of a green psychedelic revolution?

TERENCE: I think the battle is already won. No one can oppose the crusade to save the Earth; they can only quibble with its methods or style or rhetoric. No one can stand up and say, "I'm against saving the Earth."

RALPH: Lip service is not enough. People have to act collectively and according to a strategy that has a possibility of success.

TERENCE: If your enemies give lip service to your ideals, it's a signal that you are in the top position. Everybody's trying to out-green everybody else.

RALPH: The Earth's environment is a huge system, and it's not enough to recycle plastic bottles. It's not enough to stop cutting down the Amazon jungle. These things only earn a slight extension of the time available to evolve a better strategy for things like the population explosion and the exhaustion of our resources.

TERENCE: If Gaia is on our side, what fears do we need to entertain?

RALPH: The magnitude of the problem guarantees a response.

RUPERT: The green movement, if it's to be effective, must have a spiritual and mystical dimension. How does it acquire one without either allying itself to a green form of Christianity and Judaism or inventing its own kind of priestess or priest cult and carrying out its own rituals?

RALPH: There's a time-scale problem. The environmental crisis is coming upon us with the speed of a tidal wave. In the long run, it will be compared to the Flood of Noah. People had better get moving in the revitalized green church movement if they want to be on the bandwagon at all.

TERENCE: The people who can lead the psychedelicized green movement have been training themselves for years without understanding what they have been training for. Called upon to do so, they will step forward.

RUPERT: There would be an ecumenical psychedelic green order. The

Christian greens would link up with the green order of Judaism because they would have to work closely together.

RALPH: There need to be green mystical orders associated with every religion. They would serve as a kind of neural network connecting all of these diverse systems into a new unity.

RUPERT: There'd be green orders of Islam, green orders of Hinduism . . .

RALPH: . . . and Neosufism and Neokabbalah and . . .

RUPERT: The green order in America would have as one of its roles helping people reconnect with sacred places in America and honor them through appropriate ceremonies. The green order in England would have the role of connecting people with England's sacred places.

RALPH: Just as organized religion needs revolution and reinterpretation, so science also needs revolution and reinterpretation. The Gaia hypothesis and its accompanying paradigm shift in science are bringing together scientists of different specialties who have never spoken with each other before. This represents a major restructuring and synthesis of science.

TERENCE: Science for the first time has the capacity to measure its own impact on the world. Science created all these problems and science is now revealing their magnitude as they bear down upon us.

RALPH: The accommodation of a scientific view of history and archaeology by the church has to be matched by an equal resacralization of science.

TERENCE: The purpose of science should be understanding, not only technique. We need to hold back while we assimilate what we know at our current level, not push relentlessly deeper into the application of techniques. That is rape. That is violation.

RALPH: The revitalization of the church and the revision of science are already under way. In order to nucleate the social transformation implied, there must be a nucleation site, a first exemplar of this coalition of church and science. From this nucleus, the transformation might spread outward in rapid diffusion, amplified by the media.

TERENCE: A psychedelic green party.

RUPERT: The Pacific Northwest of America is one part of the world where there is an attempt to reintegrate green politics and psychedelic cults with the Judaic and Christian traditions. In Europe, lots of people use native psilocybin mushrooms, but I don't know whether they're used in a ceremonial setting. In America, mushroom cults have developed under the influence of peyote circles and Native American traditions of the sacramental use of plants. I don't think there are any living, indigenous traditions of this type left in Europe.

   The modern *ayahuasca* cult originated in the Amazon when a Christian took this substance and had a vision of Mary, who appeared to him as our Lady of the Forest. She was clad in green and revealed the outlines of the ritual use of *ayahuasca*—which the devotees call *daime*—as a communion. Such things are revealed rather than invented. They have to be channeled. If a mushroom cult were to grow up in Britain, it would have to happen spontaneously through prayer and visionary guidance.

TERENCE: The Mexican mushroom religion does its rituals in a very logical way—in small circles at night, with intentionality. There's song, there's prayer, and there's silence.

RUPERT: It's very easy to conceive of such rituals. But to have authenticity for those who participate in them, they have to be revealed rather than invented.

TERENCE: They are psychedelic experiences. Their authenticity comes from themselves. We're not talking about reciting mantras. This is the real thing.

RUPERT: Such ceremonies would need to be rooted in the spirit of the place where they occurred, and there would have to be appropriate intention and ritual elements to make them part of a larger movement such as the green order. Psychedelic groups would be the visionary branch of a larger green movement.

TERENCE: I'm proposing in my book *Food of the Gods* that the mushroom was the midwife of humanity and that all human beings in all times and places can actually claim it as their heritage.

RALPH: The ritual use of mushrooms would have been a gigantic chreode. It would be possible to excavate it at any time, anywhere.

TERENCE: Use of psychedelic mushrooms was the religion of human beings in Africa for the first million years. It underwent a retraction ten thousand years ago and broke up as recently as seven or eight thousand years ago when the progressive desertification of the Sahara changed it from a vast grassland into the formidable desert that exists today. The archaic revival seeks cultural restoration of this lost symbiotic partner. It isn't ultimately a matter of the psychedelic experience per se: psilocybin has some unique relationship to the evolution of the human nervous system. In fact, it turns the human nervous system into an antenna for the Gaian mind, assisting people to behave appropriately in the same way that termites behave appropriately within the morphogenetic field of their termite nest. If this antenna is not present in human beings, then human beings have to think up their own program, and it's usually power crazed, lethal, shortsighted, and grabby.

RUPERT: If this is the case, is your impression of people belonging to mushroom cults that they are of an entirely different quality from those who don't belong to such groups? Do they behave appropriately and in tune with the Gaian mind?

TERENCE: They tend to be rural. They tend to live communally. They tend to be nonmotivated in the economic realm. In other words, they live simply. Voluntary simplicity is a concept they're very familiar with. They love and value their children. They exemplify the values that peasantry has always exemplified because they live near the land. They want for nothing, but they have very little. This is my impression.

RALPH: Do they have a plan to take over the Christian churches?

TERENCE: No, they are beyond this. They are in the thrall of their religious relationship to the mushroom.

RALPH: They may be absorbed by this political program sooner or later, as the acid rain destroys their mushrooms.

TERENCE: My audiences until very recently preferred me to dwell on the mystical, the transhistorical. About a year and a half ago, people started questioning why there was no political content to my talks. Now they demand political content. We may imagine, because we're intellectuals and because we deal with this data, that Gaian politics is more on our minds than on the mind of the average person, but the housewife doing her ironing and the schoolchild on the bus both worry about the fate of the Earth. The collective information field has shifted its attention. The only competition for this focus on the need to save the Earth is the stupid anti-drug hysteria. It's an issue of how we relate to the vegetable matrix.

RUPERT: The psychedelic order is only one aspect of the resacralization program. Another aspect is the revival of pilgrimage and the sense of sacred time. These can be incorporated easily into many contexts of people's lives. The reason pilgrims and tourists go to sacred places is that some quality of the place is special. Rather than visiting just for historical or archaeological reasons, people can relate to the quality of a place consciously, asking the spirit of the place to inform and illuminate them on their visit and to give them its blessing. I think most people go to sacred places or into the wilderness because of a desire to make some such connection. This is part of the romantic, private, subjective tradition of our culture. It may be fairly easy to resacralize tourism. Through the impetus of the green movement, the revival of pilgrimage could spread very rapidly.

RALPH: I think we've arrived at a vision characterized by surprising optimism about the outcome of this crisis. This vision incorporates the successful escape from the downward spiral via the green movement, resacralization, and psychedelic orders. We're talking about going up the down staircase, and the way this has emerged in our trialogue, it sounds fairly plausible, possible, and underway. Perhaps it'll be the responsibility of the sacred artist to provide the links between politics, theology, and practical action.

TERENCE: Maybe what we need is a conference linking green politics and the psychedelic dimension. People who have thought a lot about green politics but not at all or mostly negatively about transformational options would have a chance to discuss these options.

RALPH: It would be a conference of synergy between the new politics and the new religion.

RUPERT: I don't think you want a conference. I think what you want is a sacrament.

TERENCE: A pilgrimage *and* a sacrament.

I think it is of primary importance to recognize consciously that education is a form of initiation. . . . In fact, modern education involves an initiation into the rationalist or humanist worldview. It elevates the intellect to a disembodied point of view in which every-thing is seen as if from the outside. . . . An alternative educational model would still be based on initiation, but a broader kind, not confined to the intellectual realm.
   —Rupert Sheldrake

In reformed education, people must be taught that history is a system of interlocking resonances in which we are all imbedded. We must teach our children that they are going to be called upon to make decisions that will affect the state of life on this planet millennia in the future.
   —Terence McKenna

The educational system of the new world order also needs the participation of the community in the determination of the curri-culum. It needs to resist evolution that's too fast while not being too rigid to change. It needs to involve a partnership of the special and the general. It needs to relate to life, not only in terms of fixing the faucets but in terms of making everyday moral decisions about altruism and selflessness and synergy.
   —Ralph Abraham

9

EDUCATION IN THE
NEW WORLD ORDER

RUPERT: Everybody agrees, even mainstream educationalists, that there's something wrong with the educational system we have today. Every society or civilization has an educational system of some kind. What would the educational system in the new world order look like?

I think it is of primary importance to recognize consciously that education is a form of initiation. Even in the present system, we have a training period and then we pass through a time of testing or trial. Some of us fail, others pass, and the passed ones become the initiates. At every level we have examinations, and each level of initiation is accompanied by impressive public graduation ceremonies. In this realm, the medieval hierarchy lives on, complete with robes, B.A., M.A., Ph.D., and so on. The initiates are like a secular priesthood qualified to run and order society. From their ranks are drawn our bureaucrats, scientists, technocrats, and intellectuals.

With certification of higher levels of education, people get better jobs, better employment opportunities, and more respect. For this reason, to the despair of educators throughout the world, most students passing through universities seem to have more interest in receiving degrees than real interest in the subjects they're studying. In the Third World, a B.A. or an M.A. changes a person's entire social status. In India, a person's marriage prospects and the size of dowry they can command depend on their degree.

In fact, modern education involves an initiation into the rationalist or humanist worldview. It elevates the intellect to a disembodied point of view in which everything is seen as if from the outside. Its slogan is objectivity. When school children are taught literature within this framework, the teacher does not read them great poems accompanied by the beating of a drum and the bringing in of magic and the realm of myth. Instead, the teacher tells them, "This poem was written by so-and-

so who was born in so-and-so and influenced by so-and-so." Students
learn facts about the poems rather than the poems themselves. This
education system makes the supreme arbiter a kind of emotionless,
detached, disembodied mind working through the medium of
written language. It tests students solely in the written rather
than the spoken mode.

The first step in this system is literacy. People must read and write
so they can know what's in reports, official documents, newspapers, and
books. This becomes more important than what they actually feel or
experience. The great libraries are like temples, containing much more
than any one of us could ever read or know about. The deeper one's
initiation into the priesthood of the written word, the less the realm of
personal experience counts for anything—except in the realm of private
life, behind the diaphragm that separates the private person from the
educated public persona.

An alternative educational model would still be based on initiation,
but a broader kind not confined to the intellectual realm. Throughout
the world, people realize that being initiated means taking on a new
social role, and usually these roles are in some sense sacralized. There
are guilds and castes of craftsmen—in India, the potter caste, the weaver
caste, the priest caste, and so on—with their own traditions and skills
passed from parents to children. Children are generally initiated into the
skills of their parents. In our society, for example, most adults want to be
initiated into the club of qualified drivers. They go through a learning
period and pass a test, and then a new freedom opens up. There is real
power with a magnetic pull and glamor to it. There are also initiations into skills like swimming and football, trades like plumbing, and
the various professions.

Much of the present educational system could be transformed if we
consciously recognized its initiatory quality. For example, medical
students, in order to become doctors, are required to dissect a human
corpse. To overcome their instinctive, deep-down revulsion to the array
of dead bodies, as well as traditional taboos against interfering with
corpses, students first entering the dissecting room presently adopt a
highly detached and usually jocular attitude. In the new system, medical
students would still dissect a human body, but they would prepare them-

selves with a meditation on death. As in some Tantric traditions where practitioners spend a night alone in a graveyard, this would make explicit the initiatory quality of the solemn moment of confronting death.

Other trades and professions would have comparable initiatory elements. Computer modeling, for example, would be an important part of the initiation into mathematics. This would introduce the initiate to the mathematical landscape, which mathematicians don't talk much about. Rather than pretending that mathematics is only a rational system of numbers and symbols, the initiate would be exposed to the vivid visual imagination, which is where creative mathematicians realize the magic is.

Included in the new educational system would be rites of passage at puberty. These could happen at new-style summer camps where there would be a program involving a vision quest, for example, with at least twenty-four hours spent alone in the wilderness. Such camps already exist in places like northern Vancouver Island, mostly for Native American youth.

One kind of system that already has this initiatory quality is the workshop system. This is at present the principal model for an alternative educational system that could replace the present one. Workshops make the dynamics of people interacting as a group explicit, and they are based on learning through experience. They attract people who actually want to learn something with others, find some new insight, or make some new transition.

TERENCE: You put your finger on the fact that the initiatory ritual is the continuing thread from the archaic that can lead us into the future. The only thing I would add is that the education of the future should have a tremendous focus on history. The educational system currently in place has as its paradigm the teaching of physics—in other words, the conveying of an extremely abstract, mathematically based description of nature that leads to high engineering competence. In an ideal educational milieu, the science of archaeology might replace the science of physics as the place where the focus is put. With the revolution in data recovery ability that has occurred in archaeology in the last ten years, a kind of telescope into the past is being erected by the world archaeology community. To teach this in our schools would release us from the post-

industrial notion of history as a kind of trendless fluctuation or class
struggle or some other very dreary model of the human journey
through time.

We've fallen into a sort of historical amnesia that has blunted the
acuity of our political decision making. In reformed education, people
must be taught that history is a system of interlocking resonances in
which we are all imbedded. We must teach our children that they are
going to be called upon to make decisions that will affect the state of life
on this planet millennia in the future. Without some knowledge of history from the birth of the universe down to yesterday's headlines, we're
not in a position to act in our own best interest. I define education
broadly as the inculcation of attitudes that cause us to act generally in
the interest of all.

RALPH: The workshop mode would be valuable to diffuse the curriculum with new dimensions of history, archaeology, and the revisioning of
the past. There could be a different teacher every year, rather than
a professor with tenure.

We've focused on higher education, and maybe when higher
education is transformed, it will somehow change the whole educational
system. I feel, however, that we haven't really addressed the main problems in the current system. The trouble is, we don't know exactly
what they are.

Infusing the current educational system with a new spirit probably
will not be sufficient. Who will be organizing all of this? Where is the
department of administration, the administration building? Who's
deciding which workshops will be offered, which teacher will conduct
them, and so on? Will there be a new emphasis on a feminist revision of
history? Will there be new interpretations of data from archaeology?
Somebody must decide, whether it's the PTA or whatever, how many
people are going to school. Everyone? A few? Those who wish? What
rewards will be offered? These things are the nuts and bolts of running
a school system. As the system evolves, or devolves, the path must be
determined and put in place at the beginning.

In the current initiation system, there are two different processes:
the initiation process and the accreditation process. This means that there

is instruction and then there is a test. As a teacher, I've always hated
the testing aspect. I am happy to teach people who want to learn from
me. That is a role I can accept. Nevertheless, I must also write letters of
recommendation or declare that a student has reached a certain level.
Usually, I don't even know what level—using the grades of A, B, C, or
whatever—the student has reached, and whatever I once thought the
dividing line between the grades was, I'm no longer sure. I like the
initiating and I don't like the testing. Yet if there is no testing, the
educational system fails in its mission.

Spiritual, moral, and social values are consistent with initiation
but not with testing. One of the things society asks education to do is
produce people qualified for trades and professions. I think the heart of a
school system's curriculum should transcend the trades, the professions,
the basic skills. Where is the spiritual initiation? Where are the moral
and ethical values? Where is the fabric of society, as it were? Where is this
to be taught, if not in the schools? Did Plato's Academy have a final exam
on Socrates' Philosophy 101?

Perhaps, in a new system, there would be a spiritual elite and profes-
sors of moral philosophy. Plato and Socrates would lead a workshop,
or something like that. The administrators arranging the plumbing workshops and those arranging the spiritual workshops would be people
with different qualifications.

Another thing that has crippled the modern university is the isolation
of the specialties. Besides having workshops with one leader, we must
have trialogue workshops to give the interplay of different specialties due
time. I don't propose, as some have, that we completely replace courses in
specialties with interdisciplinary courses. I think we need a partnership,
like Mother Earth and Father Sky. There should be time for the specialties and equal time for the syncretists to free associate, to relate the subject
matter of the entire educational experience to the progress and future
of society and to the evolutionary challenges facing each generation.

The educational system of the new world order also needs the participation of the community in the determination of the curriculum.
It needs to resist evolution that's too fast while not being too rigid to
change. It needs to involve a partnership of the special and the general.
It needs to relate to life, not only in terms of fixing the faucets but in

terms of making everyday moral decisions about altruism and selflessness and synergy.

RUPERT: Obviously, the element lacking in what I've proposed is the spiritual dimension. I took for granted, based on the present system, that the educational system is essentially secular. If we think of a spiritually based system, there's a completely different realm of possibilities. Our problem is that if the education system were Christian, then Jews, Muslims, Hindus, Buddhists, and atheists would object. This is why secularism is an important feature of modern political ideology and why spiritual traditions and practices have no place in schools. The secular state by its very nature is desacralized. It's a humanist concept.

For an entire society to have a spiritual dimension, one would need official state rituals like they have in Japan with the emperor and the Shinto religion and in Britain with the monarchy and the established church. The American model is entirely desacralized.

In America, the system would have to be a free-for-all. It would work something like this: Each student at age eighteen would be given books of fifty-five or so workshop vouchers, and the student would have to take fifty-five workshops over the next three years in order to become an initiated adult. There would have to be a minimum number taken in specific areas, like group dynamics, myths, history, philosophy, natural history, ecology, morality, and religion. To facilitate choice and logistics, there would be a computerized catalog of all the relevant programs offered at all recognized workshop centers.

RALPH: The graduating credential on finishing the use of the fifty-five vouchers would be the list of workshops completed at the point of initiation.

RUPERT: The whole process would start with a ceremonial induction into the pathway. Each workshop would itself have an initiatory pattern. The whole thing would culminate in some final test involving not only intellectual and practical skills but skills in groups as well. The final ceremony could also involve, as the Eleusinian mysteries, a psychedelic initiation, perhaps with mushrooms.

TERENCE: This would be the archaic return: a culmination of the educational process and the archaic mystery. Such ceremonies and sacraments were the original source of community. This concept follows very firmly in the steps of Aldous Huxley. In his last work, *Island*, he suggested annual ritualized encounters with psilocybin in a context of other radical forms of physical and mental therapy as a basis for a new form of education. As we reinvent Eleusis, we truly reinvent the wheel.

RALPH: This educational concept is actually a covert plan for the introduction of a new world religion through the religious aspects of initiation—and perhaps through visits to sacred sites. In the absence of an actual schoolhouse, there would be workshop centers all over the map. Workshops in religion, ethics, and so on would correspond to different established traditions as well as extinct, ancient traditions.

RUPERT: These workshops would be led by people from all traditions: Roman Catholic, Methodist, Islamic, Hindu, Tibetan Buddhist, and so on. If students wanted to know about Judaism, they wouldn't go to a professor of comparative religion, they would go to a workshop with a rabbi.

RALPH: The relationship between education and the job market would be clear in the classified ads: "Must have three W courses, two E courses, and one S course." As requirements of different industries become known, people seeking a particular profession would see to it that they took courses in the required subject areas.

RUPERT: This system could fulfill the needs of employers better than the present one.

RALPH: This vision becomes more satisfactory and plausible. What about the path that goes from here to there? How can we get rid of the entrenched system? The voters would have to have a plebiscite in order to vote on the opportunity to have a voucher system in the schools.

RUPERT: The system can simply be privatized. Vouchers would be valid at schools on an approved list drawn up by a new kind of educational board. The system would be pluralistic and extremely responsive to what people actually want and what students and parents are actually interested in. It would be decentralized and self-regulating.

TERENCE: This is school as business. Do we really want a marketplace of ideas?

RUPERT: It would be run not for profit, but through charitable trusts.

We also need to consider the reform of existing professions. Each branch of the present educational system already exists as a kind of guild of mathematicians or biochemists or engineers—with its own founding fathers, honored traditions, and so on. Each of these guilds needs to develop from within itself a vision of itself in the new world order. Groups of doctors or astronomers or geologists could get together in workshop gatherings in places like Esalen to discuss their original vision in becoming doctors. They would ask questions such as: What inspired us to study medicine? What is our present experience of the profession? What are the main limitations? What would a new vision of the healing profession be like? What could astronomy be for people today? How can the geologic profession reconnect with its patroness, Mother Earth?

RALPH: Even when the plebiscite is put on the ballot and passed, and when students are issued vouchers and the new structure simply begins, most teachers will still be ignorant of the meaning of ancient sites, the significance of stars, and the new vision of healing. Chances are they would continue teaching exactly as they teach today. While the new schools envision a new curriculum, the workshops would keep on teaching the old one.

What kind of miracle would get the whole system onto a new track? Particularly, how could the resacralization aspect that we are longing for ever happen?

RUPERT: You wouldn't just get a booklet of vouchers through the mail. When you start on your path, you would be entering a kind of apprenticeship. There would be an induction ceremony, which could be Christian, Jewish, Muslim, Buddhist, or secular. There would be plenty of scope for free enterprise in this area. You'd go through one of these ceremonies, calling in blessings on your journey. You'd get your book of vouchers as part of the ceremony. All of this would take place at a sacred place of your choosing.

RALPH: Setting up a system of initiation rites would be the key step for switching the whole system over.

RUPERT: This may become fairly easy as the importance of initiations and rites of passage for personal development becomes widely recognized.

RALPH: What about the five million young Americans who come of age for their first initiation this coming fall? Exactly how do we accommodate this number of students and produce thousands of new teachers to teach a quarter million workshops in a year?

RUPERT: Instead of an overnight change in the entire American system, I'm thinking of a pioneering experiment in a limited area.

RALPH: And it may slowly grow if it deserves to.

RUPERT: Things happen organically in society. We can't convert a system without some example of an alternative that actually works.

TERENCE: We need some concrete proof of concept demonstration.

RUPERT: The workshop system is already up and running as a concrete alternative. It exists in a pluralistic, free-market form, and it is self-sustaining. People go to workshops because they want to. Unfortunately, at present practically no one under thirty goes to workshops. It's a system of education entirely for the middle aged.

RALPH: What would be necessary to attract an eighteen-year-old to a workshop?

RUPERT: The fact is, a lot of teenagers don't know that this world of workshops exists. If they came to Esalen, for example, on an initiatory program, they would be initiated into an adult world, and new possibilities would open up to them.

Obviously, there would need to be a whole new breed of workshop leaders. Existing workshop centers would take on the new role of creating workshops to train and initiate workshop leaders.

RALPH: Somehow, standards would have to be maintained. It would have to be more than the popularity of a given workshop that guaranteed its continuous existence. Given the corruption that's a known mechanism in the downward spiral of societies, worse and worse workshops would become more and more popular because they would offer the valuation, the accreditation, and the initiation without the student actually doing anything other than sitting in a hot bath and meditating.

TERENCE: What you're implying is what you seek to avoid: a second overseeing entity that tests the workshop graduate.

RALPH: It could be that industries wouldn't employ somebody just for having graduated. They would insist on a prospective employee having gone to workshops from some of their favorite teachers or institutions. A bachelor's degree from Esalen could be worth more than a bachelor's degree from Stanford.

TERENCE: Corporations could post a list of courses that would enhance people's likelihood of being hired by them. Then students could choose for themselves which ones they would include as they formed their curriculum.

RALPH: This would be an intrinsic, self-organizational model. Maybe the corporations could just do their own testing.

RUPERT: This could start right away on a limited scale. Scholarships could be offered for, say, five workshop vouchers, with beginning and ending ceremonies for the whole thing. College students could do the workshops during their vacations. When they return to their college, they could tell their friends, who might then want to be initiated themselves. This would begin to establish a parallel system of education operating alongside the present one. If it becomes a sufficiently powerful attractor, it would have an enormous impact on schools and colleges. The traditional system's faults would become more and more apparent, because more and more people within it would have another take on what education could be like. Sooner or later, colleges would start offering workshop-type education themselves and eventually convert to the initiatory model.

RALPH: Present colleges could be persuaded to offer transfer credit for a set of five workshops in this program. Five workshops would count as one course. It would be like taking an extension course or a work/study program. In the summer, students could take workshops at five different workshop sites or just one, and they would get a college credit for it.

In sum, the new world system could actually begin with an educational program. We have to find a way to actually begin the pilot project. Since the adult education division is already functioning, we need to expand it into high school and college divisions. A few people presently coming of age, perhaps children of people who are already participants in adult workshops, could enthusiastically volunteer to be the first students of the new system. As this attractor grows and evolves and self-organizes, a bifurcation point will be reached, at which time a popular referendum could pass legislation for the fifty-five-coupon books. Existing universities and professional schools would then begin to accept workshops for transfer credit.

This system has to begin in an existing workshop center. Possibly it could be here at Esalen, since we're here dreaming this up. Assuming that corruption doesn't somehow annihilate the system as soon as it starts, it's a devious way of achieving the resacralization of the world.

*When encountered outside the religious framework, the apocalyptic
expectation of imminent transformation of the environment—with
the individual somehow playing a central role—is labeled pathology.
This pathological symptom in individuals is the driving force
behind much of our civilization.*
    —Terence McKenna

*Much more likely than any of these things we've discussed is the fact
that we're facing a serious ecological crisis and evolutionary challenge
of unprecedented magnitude.*
    —Ralph Abraham

*The attractor beyond all the doom may be another state of being
that is extraordinarily blissful compared with anything we know
here, as well as more perfect. This is the fantasy of the recovery of
Eden, the Promised Land. There's something quite magical and
infinitely attractive about this idea that has motivated the
entire historical process.*
    —Rupert Sheldrake

10

THE APOCALYPSE

TERENCE: As we near the end of our trialogues, it is fitting that we cast our minds toward final things. This seems to be not only the theme of the crisis in the present moment but the unique unifying thread throughout Western religion. More insistently than all other religious systems on Earth, the Western systems insist on appointing an end to their world. The cyclical worlds of Hinduism are cycles of time so vast that they lose all force on the popular imagination. What uniquely distinguishes Judaism, Christianity, and Islam is the insistence that God will appear tangential to history in a way that will create a last-days scenario. This scenario will involve a great uptaking of souls into the mystery of God. This idea, which is called "apocalyptic" in its more catastrophic version and "millenarian" in its more pastoral version, is a necessary correlative to the concept of Eden and the unique moment of humanity's creation by God. If human creation occurred at a unique moment in the history of the universe, then presumably, after the expiation of the sin of Eden, God will gather humans once again into the mystery.

When encountered outside the religious framework, the apocalyptic expectation of imminent transformation of the environment—with the individual somehow playing a central role—is labeled pathology. This pathological symptom in individuals is the driving force behind much of our civilization. Some years ago, the secretary of the interior was asked why he wasn't saving more of America's forests. He replied that he saw no reason to save the forests since Jesus was coming and the end of the world was imminent.

What is this intuition about the end of the world? Now that we're beginning to gather more data, science is beginning to pay back on promises made in the eighteenth century to give us a complete and deep description of the physical and astronomical universe. What we're seeing in this description is a highly chaotic domain. There isn't a stable body in the Solar System that isn't deeply pockmarked with asteroidal impact. From the inner planets to the moons of the gas giants there is tremendous visual evidence of catastrophic episodes throughout the Solar System's history.

If God or a supertransmundane event were to enter the ordinary biological and evolutionary history of a planet, there might be some kind of shock wave of anticipation, a sense of the disruption of ordinary events before the big event was in fact imminent. The brief period we've experienced of the past twenty thousand years seems to be within such an aura of anticipation. If we could strip the provincialism from the messages of the apocalyptic religions, we would find they have a deep intuition about the inherent instability of the cosmos. I think they are trying to extract something out of the human future that may in fact involve the survival of the planet. Shamans and mystics and psychedelic travelers may be getting a very noisy, low-grade signal about a future event that is somehow built into the structure of space and time.

Perhaps we are somehow witnesses to a major phase transition in the career of self-reflecting Bios in the universe. It may not be the end of the world but a complete systemic reorganization on the scale of the metamorphosis that occurs in butterflies: a complete meltdown of the previous world system and then a recasting at the behest of a higher, Gaian mind or the world soul.

Life has a terrifying tenaciousness. It seized hold of this planet 3.5 billion years ago and has managed it through hellfire. It has again and again brought the planet into stable equilibria supportive of biology. There have been asteroid infalls and continents ground to dust, and still, life has kept hold of its chunk of ground. Perhaps the recent advent of human intelligence signals a crisis of greater magnitude. The presence of our minds may indicate that we are very near some sort of enormous concrescing singularity.

We are experiencing more than calendrical pressure from the approach of the third millennium. Concrescence appears in graphs of resources and population density and demand for hydroelectric energy and levels of strontium 90 in milk. Who can look at all this data and not see either the yawning grave of humanity or a complete system reversal? I think we're standing on the cusp of a hyperdimensional event of some sort toward which all of history is being poured at a great rate. It's seeping into our religions, our politics, our dreams, and into the general imagination. We have the peculiar good fortune of fulfilling the wish conveyed in the Irish toast "May you be alive at the end of the world."

RALPH: It seems logical that if there was a Big Bang at the beginning, then there will be a Big Bang at the end. The Creation story in Genesis and the preceding myths somehow imply this.

TERENCE: In the Revelation of Saint John the Divine, this is laid out. Angels come and pour down diseases. There are plagues of scorpionlike creatures that come from the interior of the Earth. At the end, the twelve-gated city, the new Jerusalem, comes down like a flying saucer covered with jewels. It is God's kingdom coming to Earth to receive the elect as the oceans boil away and the damned are dragged into hell for eternity. This is a completely bizarre production that is one of the most puzzling pieces of literature in the Christian canon. People who immediately followed Christ expected it within their lifetime. There was a 120-year period when no one got seriously organized because everybody was standing around waiting for the end of the world. "Amen, Amen, I say unto you, this generation shall not pass away until these things are accomplished."

After about a century of this, people like Origen and Eusebius came forward and said, "Listen, enough of this waiting for the end of the world. We have to get the scene organized and get our hands on some real estate."

RALPH: I would think that the force of the prophecy has declined.

TERENCE: There are fundamentalist cults in the United States, whose adherents number in the tens of millions, that believe the scenario of Revelation is being played out on the front pages of the newspaper. The most extreme position holds that the apocalypse occurred in 1847 and that we are now living in the millennium. Then there are those who locate the date in the near future, around the year 2000 and the turning of the century. In millenarian speculation, people who discover the key invariably find the millenarian date to be just a few years ahead of their own time period.

However, if we're trying to count all the adherents to the apocalypse theory, then we have to count all the folks into scientism with their greenhouse effects, ozone holes, CFCs, and acid rain. They, too, preach apocalypse.

RUPERT: We've had the end-of-the-world scenario of nuclear holocaust hanging over us for decades.

RALPH: It comes from the same culture. I just wonder how much credibility to give the Christian apocalyptic vision when less than half the human population is involved with it. Does this mean that just half of the planet would be vaporized while the other half would keep on going?

TERENCE: All of these scenarios may be metaphors for something really weird that we are in a much better position to anticipate than John the Divine of Patmos was. In one of the scenarios I've imagined, time travel will be discovered and history will end, suddenly—just bang. People beyond that point will look back at us the way we look at the Anasazi and talk about how people used to live in linear time: all that waiting for stuff to happen in a strange jelly of stiffened dimensionality. To people born into a time-traveling world, the previous mode of existence will be mere rumor. When they are in the future, they will be able to travel back into the past, but no further than the discovery of the first time machine. Before that moment there were no time machines, and they can't take a time machine into a universe where time machines don't exist.

Imagine that it's December 22, 2012, at the La Chorrera World Temporal Mechanics Institute. The countdown is in progress and the temponaut has been strapped into the time machine. A technician pushes the button and she sails off into the future. The interesting question is, What happens in the next moment?

RALPH: Millions of people arrive from a more populated part of the universe.

TERENCE: Yes. Millions of time machines arrive from all possible parts of the future. This historic moment is as far back as people can go in their time machines. They will say to each other, "Have you been to the edge? Have you been back and seen the Abraham machine take off?"

At this point, in considering this matter of time travel, I had a serious delusional breakthrough: I realized that, when the temponaut goes off into the future, suddenly all of the future will undergo some kind of collapse and everything will happen instantly. On Earth today, the more advanced cultures tend to influence, and finally to dominate, the less advanced cultures. This is similar to the equalization of pressure inside a closed chamber of gases. In the same way that gases confined in a space equalize pressure to a uniform value, cultures tend to take on some of the characteristics of the most advanced cultural level with which they are in contact. This happens whether the cultures in question are confined to a single planet and a single historical epoch or are confined within a temporal domain defined by the limits of a time-traveling technology.

The most advanced state of human accomplishment, even if it is billions of years in the future and absolutely beyond our ability to imagine, will appear one millisecond after the temponaut takes off, on the other side of the time threshold. This technology takes the entire future history of the universe, up until its conclusion, and compresses it down into the next few milliseconds. We will then be face to face with the end purpose of all evolution, all process, all pattern, all energy, space, time, and matter.

This is a complete fulfillment of the monotheistic intuition about the apocalypse. It is as though the universe is a huge conundrum, and we're in there suffering through a long strange trip. There's science and religion and magic, and we're fumbling and fumbling slowly toward something. It turns out to be the alchemical gold, and when we clasp it to ourselves, time ends, space ends, matter ends, and everything ends. We go into the conclusion, the payoff, the jackpot. We go over the cusp and meet the management.

RALPH: Probably, this hasn't happened yet.

TERENCE: No. It will occur in A.D. 2012. At least that's the implication I draw from my work with the fractal wave that was so carefully built into the I Ching by the pre-Chou Chinese. There is very good agreement between this wave and recorded historical data, but only if we make two assumptions. One is that the wave maps the ingression of novelty, or complexity, into three-dimensional space/time. The second assumption is that the wave's terminus is late in A.D. 2012. With these assumptions in place, there is a very neat formal mapping for the most difficult of all phenomena to model: human history itself. The fact that the Mayan calendar ends on the same date gives me some measure of confidence.

RALPH: There are alternatives in the interpretation of this date. I think that this particular time-travel fantasy of yours is actually a syncretism between apocalyptic paranoia and the time wave.

TERENCE: It isn't necessary to associate apocalypse with the date. The fact that the wave fits so well with all this millenarian and apocalyptic pressure at the turn of the millennium seems pretty suggestive to me. The Mayan calculations are another coincidence. The Maya and the Christians are within twelve years of each other if we take the year 2000 as the turn of the eon. This is the slippery realm of human judgment and data evaluation.

RALPH: Ruth Benedict studied sixty different cultures, charting them out by different parameters and finally sorting them into three bins: the Apollonian, the Dionysian, and the Paranoid. It just may be that a Paranoid culture having a religion with a paranoid element, contributed by Saint John the Divine, happens to develop extraordinary technical powers that are great generators of lethal toxicity.

TERENCE: The word *paranoid* is designed to make people not like it. If something has this label, nobody will seriously look into it. There is an implicit assumption that there's nothing to be paranoid about. In fact, in a very dynamic and unsteady universe, paranoia may well be a true sensitivity to the facts of the matter.

RALPH: Paranoids always say that. I know there are fundamentalist Christians around who take every word of the Bible very seriously and literally. However, as far as I'm concerned, the Bible is somebody's paranoid fantasy put down in a book. I certainly don't believe that it's a divine document with any special credibility. Some people have taken this book very seriously, so now they're paranoid too.

I'm ready to admit that there are a number of coincidences about this year 2012 and that some of them are ominous. But I'm still not giving any credibility to Saint John the Divine. From the morphogenetic-field point of view, there are quite a number of people believing in Saint John the Divine. That we have to take seriously.

TERENCE: He felt a quaking in the force. Today we have much better techniques than Saint John the Divine to figure out what this quaking in the force is.

RALPH: Much more likely than any of these things we've discussed is the fact that we're facing a serious ecological crisis and evolutionary challenge of unprecedented magnitude. James Lovelock has said that the present rate of species extinction is one of the eight largest catastrophes in the planet's lifetime.

TERENCE: If we don't have a miracle every day, we're not going to make it. That's why we don't need Saint John the Divine to tell us there's an apocalypse underway.

RALPH: Every prophecy with any credibility says to me that there is some way to make it through. I have to admit I'm extremely doubtful of the intelligence of this human species to find it.

TERENCE: It depends on what's causing the problem. If you think humans are the problem . . .

RALPH: Humans *are* the problem.

TERENCE: All parameters of planetary stability become more and more unstable as time approaches the present. What about the sudden appearance of large and repeated glaciations in the last five million years? Glaciers are new in the life of the Earth and may indicate that something is wrong with the sun or the geodynamics of the planet. Maybe humans are the problem. Or maybe human beings are the answer.

RALPH: The Earth could jump off its orbit at any moment and head out to space, but it seems to me that ecological catastrophe is a more appropriate form of apocalyptic vision at the present time.

RUPERT: There's a sense in which the apocalyptic scenario we find ourselves in is a product of the apocalyptic myth throughout history. It's a self-fulfilling prophecy. The apocalyptic tendency in Christianity inspired millenarian movements throughout the ages, including that of the Pilgrim

fathers, who came to a new world in America—the promised land.
It inspired Francis Bacon's vision of unlimited progress through science
and technology and the conquest of nature. His millenarian goal was a
technological utopia, a new Eden of peace, prosperity, and wise scientist-
priest figures running everything, a promised land flowing with milk and
honey and material abundance brought about through the scientific
control of nature. This scenario underlies the ideology of progress
and is now bringing about ecological catastrophe.

The myth of Faustian science is related to this apocalyptic model.
In the sixteenth and seventeenth centuries, Faust sold his soul to the devil
in return for unlimited knowledge and power for a fixed period—twenty-
four years—after which he was damned and dragged down to hell. The
modern form of the Faust myth is Mary Shelley's Frankenstein story,
in which the scientist is not destroyed by supernatural powers but by his
own creation. It's obvious that the nuclear threat has a Frankenstein
quality to it, and the ecological crisis also has this apocalyptic
mythic basis.

RALPH: For this reason, we have to do surgery on the self-fulfilling
mythological mechanism working in history. One good start would be
a reinterpretation of the Revelation of Saint John the Divine.

TERENCE: We need to switch the vision onto another track.

RUPERT: The Revelation of Saint John the Divine is not a unique
phenomenon in the Judeo-Christian tradition. Around the time of Christ,
many people believed the end was at hand. In this sense, it was a period
very similar to our own. The Book of Daniel in the Old Testament is
an apocalyptic, prophetic book and is a precursor of the Revelation
of Saint John the Divine. These are just two examples in a large and
extensive literature.

An apocalyptic spirit pervades the teachings of Jesus. Indeed, it runs
through the entire Bible. God promises Abraham that he will take him
and his descendants to another land where wonderful things will happen
and his children shall be as the sands of the sea and he shall be the father
of many nations. Through faith in such promises, history has been made.

Faith allowed Moses to lead the people of God out of Egypt into the Promised Land.

In such stories, there is a fundamental sense of being on a journey toward some wonderful destiny in the future. This journey can be to a different place, like America for the Pilgrim fathers. It can also be a journey through history to a future millennium, to a new age. This is the faith that underlies the attempt to transform the world through science and technology. It's a pattern so deeply rooted in the Judeo-Christian tradition, so fundamental to the entire historical orientation of our religion and culture, that mere tampering with the book of Revelation won't make it go away.

TERENCE: How do we direct history toward a nonlethal yet satisfying conclusion? Perhaps history isn't simply a lethal neurosis. Perhaps it's an actual anticipation of what has been made inevitable by all this technology.

RALPH: You are suggesting that we just get used to the apocalypse happening.

RUPERT: We've all got used to the fact it could happen at any time. Terence, your message recalls the saying of Jesus that the kingdom of heaven will come like a thief in the night.

TERENCE: " . . . like a thief in the night, and no man will know the moment of my coming."

RALPH: Either there is an inevitable apocalypse on the horizon or one might be created by a self-fulfilling mechanism of paranoid prophesy. To stop it, we must defuse the time bomb of the Bible.

TERENCE: We should be allowed to let the apocalypse happen, rather than make it happen, which is what we seem to be set on doing.

RALPH: The story of the secretary of the interior is a direct example. For him, not only is it happening, but it's happening so soon that we may as well kill off everything immediately for the fun of it.

TERENCE: "Things fall apart; the centre cannot hold; Mere anarchy is loosed upon the world. . . . What rough beast . . . Slouches toward

Bethlehem to be born?" This image from William Butler Yeats haunts the twentieth century as strongly as it haunted the first and second centuries A.D. If culture is a fantasy arising from the unconscious, then we've certainly set ourselves up for the end. It's going to be very delicate to ride this through, understand it, stop it, back out of it, and integrate it.

RALPH: Is this one of our major weaknesses from the evolutionary point of view?

RUPERT: It's what has enabled us to understand and discover the evolutionary point of view. The very notion of human progress is an apocalyptic vision of history written large. This myth is not confined to churches and synagogues. Our scientific worldview has grown up within the Judeo-Christian matrix with its idea of a beginning, a middle, and an end. The idea of human progress was widespread by the end of the eighteenth century, and the idea of biological progress extended the same idea to all life, giving us the theory of evolution. Since the 1960s, the entire cosmos has been seen as evolutionary. I would say that the Big Bang cosmology is an apocalyptic vision of history with an explosive beginning implying an explosive or implosive end. Astronomy points to a more local apocalypse as well. The theory is that, sooner or later, maybe in five billion years, the sun, like any other star, will burn up its hydrogen fuel, get much hotter, expand, and then puff off its outer atmosphere, leaving a white dwarf core. This will be the end of all life in the solar system. This scientific worldview is undoubtedly apocalyptic, but it puts the end in the remote future.

TERENCE: Thus conveniently far away.

RUPERT: The mechanistic science of the nineteenth century also predicted the final heat death of the universe. This would be the ultimate triumph of entropy or chaos over order.

RALPH: The apocalypse myth is an integral part of the historical concept of the Israelites who invented it, and we believe we can't do anything about it. The paranoia of our culture is manifest in the assumption that the end is happening tomorrow.

TERENCE: This is more than paranoia. The Earth is on fire, haven't you heard? Who else has nuclear stockpiles? Who else has Agent Orange? Who else has CFCs dissolving the ozone hole? We can interpret this as a slow apocalypse that takes two hundred years or as a fast apocalypse that takes fifteen minutes and can happen today or tomorrow. All these possibilities are real.

RALPH: Your projection of apocalypse in the year 2012, I think, is actually damaging our chances of having a future.

TERENCE: I think it's a way to manage ourselves through a narrow neck in a state of high awareness. We can use the calendar as a club, saying to our leaders, "Do you want to enter the third millennium armed like barbarians? Or do you want to drape yourself in the mantle of peace and be the saviors of the world, the unifiers of mankind?" Everyone should thoroughly examine the premises of their society as we approach the third millennium. For example, the approaching millennium is putting tremendous pressure on governments to get rid of nuclear stockpiles by the year 2000.

By the year 2012, the world population will be approaching ten billion people. Propagated at the present rate of fade, there will be no ozone layer. The impact of that single parameter is totally unknown. Then there are carbon dioxide emissions, acid rain, and nuclear proliferation and propaganda running rampant. Meanwhile, pharmacology, brainwashing, surgical reconstruction, and high-tech undercover technologies of all sorts are making new leaps toward their own twisted perfection. Under these conditions of cultural compression, forms of novelty will erupt that are totally unpredictable in the present context. Everything is knitting together. Our boundaries are dissolving into a kind of techno-biological-informational soup. The intentionality behind all this is in the hands of no one but the Gaian will.

What is happening is like the metamorphosis that goes on inside a chrysalis. This planet is having its forests liquified, its oceans boiled, its populations moved, and its genes are streaming in all directions with exotic toxins mixed in. We're in a timestorm whose diameter is impossible to estimate. The barometric pressure is dropping faster than we've ever seen it drop. There's an eerie stillness, and the light in the sky looks very strange, but nothing definite has happened yet.

RUPERT: A nuclear holocaust in which Christendom destroyed itself would be a self-fulfilling apocalypse. The environmental threat is much more global. As we shift the focus of our attention from the dangers of nuclear war and turn to the global problem, it is apparent that the Omega Point you are trying to describe, Terence, involves some kind of collective transition in human consciousness. This might be achieved without mass death through something you have vividly portrayed as a kind of collective hallucinogenic experience.

TERENCE: Every human life becomes ultimately an approach to this question of final time. If we don't live in the age of the world's end, that doesn't mean we don't get to deal with the question of final time. We all die. It's just that, in this age, the death of the individual and the death of the species are somehow both possible to contemplate.

RALPH: And the death of all other species.

TERENCE: Many traditions teach that life is organization for the purpose of creating a kind of after-death vehicle in a higher dimension that will survive the transition. Building such a vehicle is seen as the true purpose of life. This is the transition to light that so many traditions have anticipated.

RUPERT: What people believe happens to them after death makes a difference in the way they face death. Many people who have had a near-death experience say they no longer fear death because they know there is something beyond it. It is possible that vast numbers of people going through the barrier of death at the same time may be creating a kind of group mind of a kind never before realized.

Why the apocalypse is such a strong attractor is an interesting question. The attractor beyond all the doom may be another state of being that is extraordinarily blissful compared with anything we know here, as well as more perfect. This is the fantasy of the recovery of Eden, the Promised Land. There's something quite magical and infinitely attractive about this idea that has motivated the entire historical process.

RALPH: It is the Rapture. This is the antidote that's more or less built into the apocalyptic vision in Saint John.

RUPERT: Let me add one more ingredient to this particular line of thought. I recently took part in a discussion with Brian Swimme, who was exploring the idea that the universe, like a developing organism, has phases at which particular kinds of things happen. In an evolutionary cosmos, there is a time when atoms first come into being. Then there is a time when galaxies form. Then there is a time when the stars are old enough to explode into supernovas, releasing the stardust out of which planets are made. Like a developing embryo, cosmic development has particular phases that are roughly synchronous throughout the universe. Therefore, if there is life on other planets, its evolutionary stage might not be very different from ours.

I had also been thinking about parallel evolution on other Earth-like planets and the possible effects it would have on our own evolution through morphic resonance. I asked Brian for a rough estimate of how close the development of such other Earthlike planets might be to our own. He said perhaps within 50,000 to 500,000 years. If there is indeed morphic resonance between similar planets, then when a new form appears on Earth it's more likely to appear on other planets. If any planets got far in the lead, morphic resonance would tend to make the others catch up. With cosmic synchronization through morphic resonance, there is the sense of a possible cosmic apocalypse.

RALPH: Death on a cosmic scale.

RUPERT: Or the total transformation of the soul of the world.

RALPH: Well, I hope we can transform the apocalypse myth and make it suitable for something other than destruction.

RUPERT: In the green psychedelic churches of the Amazon, there is an Incan version of the apocalypse myth in which a dragon in the last days comes and eats up the forest, burning and destroying everything.

In the last days, the struggle between the serpent and the forces of life grows ever more intense. People are forced to take sides. It is no longer possible to sit on the fence because the fence itself is crumbling.

There will be an intense polarization as the new millennium approaches because these forces will become ever more powerful, preparing for the final battle. Through faith in victory over the dragon, victory will be achieved.

A dragon, incidentally, is prophesied at the very beginning of Anglo-Saxon liberal political theory. In Hobbes' model, individuals are like atoms in the body of Leviathan. The dragon that is destroying the Amazon forest is the great Leviathan of modern society. The struggle is going on now; the outcome is uncertain.

Rubbing out the apocalyptic model, getting rid of it, suppressing it, or psychically engineering it out of our psyche is perhaps no longer possible. The dragon of destruction, Leviathan, is itself motivated by a millenarian faith: the dream of conquering nature and subduing its destructive powers. Whereas for the heroes of scientific progress the dragon was nature, for modern greens the dragon is the human system; this Leviathan is devouring the forests, burning things up, and polluting the world.

We're in the morphic field of the millenarian process. Only another millenarian scenario can undo the earlier one that has proved so destructive.

RALPH: From chaos we came and unto chaos we shall return.

TERENCE: The middle name of chaos is opportunity.

GLOSSARY

ALCHEMY: An ancient science, closely related to astrology, developed from the interaction of Chaldean and Aristotelian theories of matter and from which chemistry evolved in British and Continental scientific societies following the Renaissance. Alchemists, who belonged to the Hermetic tradition, sought to transmute base matter or spirit into noble through magical and chemical manipulations. (Cf. *Hermetic tradition.*)

ANIMISM: The view that nature is alive rather than inanimate.

ANTIPARTICLE: A subatomic particle that has the same mass as another particle, but an equal and opposite value of some other property. For example, the antiparticle of a negatively charged electron is a positively charged positron.

APOCALYPSE: The Judeo-Christian and Islamic belief in the entry of God into history and the subsequent end of the world.

ASTRAL PLANE: According to occult doctrines, a plane of existence beyond the physical realm; the first sphere of existence after the death of the body, which also can be visited in dreams and out-of-the-body journeys.

ATTRACTOR: In the mathematical theory of dynamical systems, an irreducible invariant set that attracts the trajectories of all nearby points.

BIFURCATION THEORY: A branch of chaos theory dealing with the changes in the configuration of attractors caused by changes in the rules defining the dynamical system.

CHAOSCOPY: A computer-based technique for the observation of the hidden form within chaotic data. Also known as attractor reconstruction.

CHAOS THEORY: The branch of mathematics dealing with dynamical systems. Also known as dynamical systems theory.

CHAOTIC ATTRACTOR: Any attractor that is more complicated than a single point or a cycle.

CHREODE: A canalized pathway of change within a morphic field.

CONCRESCENCE: From the metaphysics of Alfred North Whitehead, concrescence is the knitting together of disparate elements into a unified nexus.

COSMOLOGY: The study of the evolution, general structure, and nature of the universe as a whole.

DAKINI: Literally, "sky dancer." A dynamic, energizing, and feminine principle in Tibetan tantrism. She may manifest as a human being, as a peaceful or wrathful goddess, or as the general play of energy in the phenomenal world.

DARK MATTER: Also known as the "missing mass"; seems to make up from 90 to 99 percent of the matter in the universe and is of unknown nature.

DEISM: Belief in God on the basis of reason alone, usually confining the role of God to creating the universe and establishing the laws of nature.

DETERMINISM: The doctrine that all events, including human actions, are predetermined.

DIALECTIC: The conversational method of argument, involving question and answer. In Hegelian and Marxist philosophy, a pattern of development by means of contradiction and reconciliation, involving thesis, antithesis, and synthesis.

ECOSYSTEM: A community of organisms together with the environment in which they live; for example, a tropical rain forest.

ELEMENTALS: Personified natural forces; faeries, sprites, and nixies are among the elementals.

ENTELECHY: In the philosophy of Aristotle and in vitalist biology, the principle of life, identified with the soul or psyche. Entelechy gives an organism its own internal purposes and defines the end toward which it develops.

ENTROPY: A quantity defined in terms of thermodynamics. The entropy of a system is the measure of its degree of disorder. According to the second law of thermodynamics, the entropy of closed systems increases with time.

EPIGENESIS: The origin of new structures during embryonic development.

ESCHATOLOGY: A branch of theology dealing with the "four last things": death, judgment, heaven, and hell.

EUCARYOTES: Living organisms consisting of cells with nuclei, such as fungi, plants, and animals. (Cf. *procaryotes*.)

FASTNACHT: A medieval festival, recently reinstituted to relieve the boredom of winter in Switzerland.

FIELD: A region of physical influence. In current physics, several kinds of fundamental fields are recognized: the gravitational and electromagnetic fields and the matter fields of quantum physics. In biology, morphogenetic fields (q.v.) organize the development and maintenance of bodily form. According to the hypothesis of formative causation (q.v.), morphic fields organize the structure and behavior of organisms at all levels of complexity and contain an inherent memory.

FORMATIVE CAUSATION: The hypothesis, first proposed by Rupert Sheldrake in 1981, that self-organizing systems at all levels of complexity are organized by morphic fields, which are themselves influenced and stabilized by morphic resonance from all previous similar systems.

FRACTAL: Name introduced around 1967 by Benoit B. Mandelbrot for a geometric object with a fractional dimension, such as the coastline of California near Big Sur.

GAIA: Mother Earth. The Gaia hypothesis, proposed by James Lovelock, regards the Earth as a self-regulating, living organism.

GEODYNAMICS: Continental drift, earthquakes, volcanism, atmospheric and oceanic currents, and other physical processes that shape the Earth.

GNOSTICISM: The belief in salvation or liberation through esoteric knowledge; usually associated with a sharp distinction between the spiritual world, regarded as good, and the material world, regarded as evil.

HERMENEUTICS: A philosophical tradition, evolving from Old Testament scholarship into literary criticism, cultural history, and the philosophy of science, in which the intellectual functions of perception, interpretation, and construction of consensual reality are closed into a loop called the critical circle or hermeneutical cycle.

HERMETIC TRADITION: One of the spiritual traditions of late antiquity in Alexandria, tracing its source to Hermes Trismegistus. It produced the *Hermetica*, which include the *Hermetic Corpus* (containing about seventeen books, including the *Poimander*), the *Asclepius*, and other writings. The *Hermetic Corpus* presents alchemical, magical, astrological, and philosophical doctrines of liberation. The *Poimander* (divine mind) speaks to Hermes, who receives a vision of light, from which the logos and then the universe are created. The *Asclepius* explains the creation of everything from the One, including the hierarchy of angels. The Hermetic tradition posited that the elemental world was infused by astral influences—"as above, so below." It exerted great influence on the Neoplatonists of the Florentine Renaissance such as Marsilio Ficino and Pico della Mirandola. (Cf. *alchemy, logos.*)

HOLOGRAM: A photographic record of a three-dimensional object made with a split beam of light from a laser. From a part of a hologram, an image of the whole object can be reconstituted.

HUMANISM: In its literary sense, the intellectual movement that characterized the culture of Renaissance Europe. In its usual modern sense, a rejection of all religious beliefs and an insistence that we should be exclusively concerned with human welfare in this material world, assumed to be the only one.

HYPOSTATIZATION: A concept symbolized in concrete form; the process of ascribing material existence to something.

LOGOS: A divine realm, basic to the Alexandrian tradition since the time of Philo the Jew.

M-FIELD: An abbreviation for morphogenetic field or morphic field.

MECHANISTIC THEORY: Based on the metaphor of the machine. The doctrine that all physical phenomena can be explained mechanically, without reference to goals or purposive designs. (Cf. *teleology.*)

MILLENNIUM: The year ending a thousand-year period, for example A.D. 2000. In Christian theology, the millennium refers to a thousand-year period of peace and prosperity expected to occur immediately before the end of the world.

MITOSIS: The usual process by which the nucleus of a living cell divides into two.

MORPHIC FIELD: A field within and around a self-organizing system that organizes its characteristic structure and pattern of activity. According to the hypothesis of formative causation, morphic fields contain an inherent memory transmitted from previous similar systems by morphic resonance and tend to become increasingly habitual. Morphic fields include morphogenetic, behavioral, social, cultural, and mental fields. The greater the degree of similarity, the greater the influence of morphic resonance. In general, systems most closely resemble themselves in the past and are subject to self-resonance from their own past states.

MORPHIC RESONANCE: The influence of previous structures of activity on subsequent similar structures of activity organized by morphic fields. According to the hypothesis of formative causation, morphic resonance involves the transmission of formative influences through or across time and space without a decrease due to distance or lapse of time.

MORPHOGENESIS: The coming into being of form.

MORPHOGENETIC FIELDS: Fields that play a causal role in morphogenesis. This term, first proposed in the 1920s, is now widely used by developmental biologists. According to the hypothesis of formative causation, these fields contain an inherent memory, transmitted from similar past organisms by the process of morphic resonance.

MYCELIUM: The undifferentiated threadlike tissue that precedes the development of a fruiting body in the life cycle of mushrooms.

NEOPAGAN TRADITION: A new wave of resurgence of pre-Christian religion and belief.

NEOPLATONISM: A development of Plato's philosophy combining mystical, Oriental, and Aristotelian influences, first systematized in Alexandria in the third century A.D. by Plotinus. Like Platonic philosophy, it postulates a transcendent realm of changeless archetypes or Forms; it also emphasizes that just as all living beings are animated by immanent souls, so the entire cosmos is animated by the world soul.

NOVELTY WAVE: Hypothetical alternative to probability theory developed by Terence McKenna; the novelty wave is variable, which determines the rate and times at which statistically improbable events may occur.

OMEGA POINT: The state of complex unity toward which everything is developing, according to the philosophy of the evolutionary mystic Teilhard de Chardin, who described it as "a distinct Centre radiating at the core of a system of centres . . . a supremely autonomous focus of union."

ONTOLOGY: The philosophical study of existence itself, differentiating between "real existence" and "appearance." Also, the assumptions about existence underlying any theory or system of ideas.

OVERSOUL: A synonym for the supraphysical realm called the world soul by Plato in his dialogue, the Timaeus, resurrected under this name by the American transcendentalists Emerson and Thoreau.

PARADIGM: An example or pattern. In the sense of the philosopher T.S. Kuhn, scientific paradigms are general ways of seeing the world that are shared by members of a scientific community, and they provide models of acceptable ways in which problems can be solved. Scientific revolutions are associated with changes of paradigm, or "paradigm shifts."

PARANORMAL PHENOMENA: Little-understood and often elusive phenomena that lie outside the scope of current scientific orthodoxy.

PATRILINEAL: The form of society in which property belongs to men and descends from father to son.

PHEROMONES: Organic chemicals produced to convey information among the various members of a species; ants, for example, use pheromones to communicate.

PINEAL GLAND: A small organ of uncertain function found in the human brain at the roof of the third ventricle. Descartes placed the "seat of the soul" in the pineal.

PLEBISCITE: An expression of popular will; for example, by direct vote.

PRIMAL UNIFIED FIELD: Hypothetical state of the early universe, before the symmetry break that gave rise to the four elementary forces. (See superstring theory.)

PROCARYOTES: Cells or organisms lacking cell nuclei, such as bacteria and blue-green algae. (Cf. *eucaryotes*.)

PSYCHEDELICS: A family of psychoactive indole compounds such as LSD, psilocybin, and DMT that cause visionary hallucinations.

PULSAR: A compact star produced as the remnant of a stellar explosion. Pulsars spin very rapidly, and, through magnetism and charged particles, produce very regular bursts of radio noise.

QUANTUM THEORY: A departure from classical Newtonian mechanics, based on the principle that certain physical quantities can assume only discrete values. Quantum mechanics has several seemingly paradoxical features, including the way that entities such as photons and electrons can be regarded as both waves and particles.

QUASAR: An object that appears to be an ordinary star but with very large red shifts. A quasar is a quasistellar object.

REDUCTIONISM: The doctrine that complex systems can be explained in terms of simpler ones; for example, living organisms in terms of inanimate physico-chemical processes.

RESACRALIZATION: The recognition of the sacredness of that which has been desecrated or treated as devoid of spiritual presence.

RESONANT WAVE PHENOMENON: The production of a wavelike pattern in one elastic medium through weak coupling to a similar pattern in another nearby medium by a process of resonance; for example, when one piano string is struck, a sound is excited in nearby strings.

RUNNEL: Properly, a small brook or watercourse. In morphic fields, a well-worn path or habit of thought.

SCIENTISM: A faith in natural science as the only valid source of authority.

SEED CRYSTALS: Crystals introduced into a saturated solution to cause crystallization of the dissolved material.

SEROTONIN: One of several neurotransmitters necessary to ordinary brain function.

SHAMAN: Practitioner of an archaic style of healing and natural magic. Shamans are masters of the archaic techniques of ecstasy.

SINGULARITY: In physics, a domain or situation in which the laws of physics either do not apply or have been broken down; the center of a black hole is a singularity, for example.

SPACE/TIME CONTINUUM: The four-dimensional geometric model for natural histories popularized by Albert Einstein in his theories of relativity.

SUPERCONDUCTIVITY: Zero-resistance conduction of electricity by some metals and alloys at low temperatures.

SUPERSPACE: Hypothetical dimensions in which ordinary space and time are embedded.

SUPERSTRING THEORY: First proposed in the 1980s, superstring theory models particles not as points but as vibrating and rotating "strings." In one version there are ten dimensions, nine of space and one of time. Superstring theory postulates an original unified field at the birth of the cosmos that gave rise to the known fields of physics as the universe expanded.

SYNCRETISM: The attempt to blend together seemingly inharmonious elements from different systems of philosophy or religion.

TELEOLOGY: The study of ends or final causes; the explanation of phenomena by reference to goals or purposes.

TEMPONAUT: A time traveler.

THEOSOPHY: An esoteric system of understanding the nature of the divine and its relation to the living cosmos. The Theosophical Society, founded in 1875, draws together Hindu, Buddhist, Western and other wisdom traditions, and it has done much to disseminate occult ideas in the Western world.

UROBORIC SYMBOL: Ancient image of a snake taking its tail in its mouth; symbol of eternity and the completed alchemical process.

WORLD SOUL: Also known as the *anima mundi*, the animating principle of the whole world.

BIBLIOGRAPHY

Abraham, R., and C.D. Shaw. *Dynamics, The Geometry of Behavior.* Second edition. Reading, MA: Addison-Wesley, 1992.

Argüelles, J. *The Mayan Factor: Path Beyond Technology.* Santa Fe, NM: Bear & Co., 1987.

Bergson, H. *Creative Evolution.* London: Macmillan, 1911.

Eisler, R. *The Chalice and The Blade.* San Francisco: Harper & Row, 1987.

Eliade, M. *Shamanism: Archaic Techniques of Ecstasy.* New York: Pantheon Books, 1964.

Fox, M. *The Coming of the Cosmic Christ.* New York: Harper and Row, 1988.

Gimbutas, M. *The Goddesses and Gods of Old Europe.* Berkeley, CA: University of California Press, 1982.

Graves, R. *Difficult Questions, Easy Answers.* New York: Doubleday, Inc., 1964.

_____. *Food for Centaurs.* New York: Doubleday, Inc., 1960.

_____. *The White Goddess.* New York: Creative Age Press, 1948.

Haraway, D.J. *Crystals, Fabrics, and Fields: Metaphors of Organicism in Twentieth-Century Developmental Biology.* New Haven, CT: Yale University Press, 1976.

Hoffer, A., and H. Osmond. *The Hallucinogens.* New York: Academic Press, 1967.

Huxley, A. *The Doors of Perception.* New York: Harper & Row, 1970.

_____. *Island.* New York: Harper & Row, 1972.

_____. *Moksha.* Ed. Horowitz and Palmer. Los Angeles: Jeremy P. Tarcher Inc., 1977.

Jantsch, E. *Design for Evolution.* New York: George Braziller Inc., 1975.

Jantsch, E., and C.H. Waddington. *Evolution and Consciousness.* London: Addison-Wesley, 1976.

Jantsch, E. *The Self-Organizing Universe.* New York: Pergamon Press, 1980.

Jaynes, J. *The Origin of Consciousness in the Breakdown of the Bicameral Mind.* Boston: Houghton Mifflin Co., 1977.

Jung, C.G. *Flying Saucers: A Modern Myth of Things Seen in the Sky.* Princeton, NJ: Princeton University Press, 1978.

_____. *Mysterium Coniunctionis.* New York: Pantheon Books, 1963.

_____. *Psychology and Alchemy.* London: Routledge & Kegan Paul, 1953.

Koestler, A. *The Ghost in the Machine.* New York: The Macmillan Co., 1967.

Lovelock, J. *The Ages of Gaia: A Biography of Our Living Earth.* New York: Bantam Books, 1988.

Mandelbrot, B. *The Fractal Geometry of Nature.* San Francisco: Freeman & Co., 1977.

McKenna, D., and T. McKenna. *The Invisible Landscape.* New York: Seabury Press, 1975.

McKenna, T. *The Archaic Revival.* San Francisco: Harper San Francisco, 1992.

_____. *Food of the Gods.* New York: Bantam, 1992.

Munn, H. "The Mushrooms of Language." In *Shamanism and Hallucinogens,* edited by M. Harner. London: Oxford University Press, 1973.

Oss, O.T., and O.N. Oeric. *Psilocybin: Magic Mushroom Grower's Guide.* Berkeley, CA: Lux Natura, 1985.

Otto, W.F. *Dionysus: Myth and Cult.* Bloomington: Indiana University Press, 1965.

Prigogine, I. *From Being to Becoming.* San Francisco: Freeman & Co., 1980.

Prigogine, I., and G. Nicolis. *Self-Organization in Nonequilibrium Systems.* New York: John Wiley & Sons, 1977.

Schultes, R.E., and A. Hofmann. *The Botany and Chemistry of Hallucinogens.* Springfield: Thomas, 1973.

_____. *Plants of the Gods.* New York: Alfred van der Marck, 1979.

Sheldrake, R. *A New Science of Life.* Los Angeles: Jeremy P. Tarcher, Inc., 1982; new edition 1988.

_____. *The Presence of the Past.* New York: Times Books, 1988.

_____. *The Rebirth of Nature.* New York: Bantam Books, 1991.

Smith, P. *The Killing of the Spirit.* New York: Viking Press, 1990.

Stewart, I. *Does God Play Dice?* Oxford: Basil Blackwell, 1989.

Thom, R. *Structural Stability and Morphogenesis.* Reading, MA: Addison-Wesley, 1975.

Waddington, C.H. *The Nature of Life*. London: George Allen & Unwin Ltd., 1961.

_____. *The Strategy of the Genes*. London: Allen and Unwin, 1957.

Waite, A.E. *Book of Ceremonial Magic*. Secaucus, NJ: Citadel Press, 1970.

Wasson, R.G. *Soma: Divine Mushroom of Immortality*. New York: Harcourt Brace Jovanovich, 1971.

Wasson, R.G., A. Hoffman, and C. Ruck. *The Road to Eleusis*. New York: Harcourt Brace Jovanovich, 1978.

Wasson, R.G. *Persephone's Quest: Entheogens and the Origins of Religion*. New Haven: Yale University Press, 1986.

Whitehead, A.N. *Process and Reality*. New York: Macmillan Co., 1929.

Wolfson, H. A. *Philo*. 2 vols. Cambridge, MA: Harvard University Press, 1947.

ABOUT THE AUTHORS

RALPH ABRAHAM

Ralph Abraham was born in Vermont in 1936 and earned a Ph.D. in mathematics from the University of Michigan in 1960. He participated in the creation of global analysis, a new branch of mathematics, while teaching at the University of California at Berkeley, Columbia University, and Princeton University. He has been at the University of California at Santa Cruz since 1968, where he has been a leader in the new theories of nonlinear dynamics, chaos, and bifurcation.

Abraham is the author of several mathematical texts, including the pictorial introduction to dynamics, *Dynamics, the Geometry of Behavior*. After a three-year tour of Europe and India, he began a program of expanding mathematics into its role in individual and social evolution, manifest today in a series of articles and a book, *Chaos, Gaia, Eros*. He lives in a redwood forest and frequently goes skiing and surfing with his two grown sons.

TERENCE McKENNA

Born in 1946, author and explorer Terence McKenna has spent the last twenty-five years studying the ontological foundations of shamanism and the ethnopharmacology of spiritual transformation. He graduated from the University of California at Berkeley with a distributed major in ecology, resource conservation, and shamanism. After graduation, he traveled extensively in the Asian and New World tropics, becoming specialized in the shamanism and ethno-medicine of the Amazon Basin.

With his brother Dennis, McKenna is the author of The *Invisible Landscape* and *Psilocybin: The Magic Mushroom Growers' Guide*. A talking book of his Amazon adventures, *True Hallucinations*, has also been produced. His two most recent books are *Food of the Gods* and a book of essays, *The Archaic Revival*.

He is the father of two children, Finn and Klea, and currently lives in California and Hawaii. He is a founder and director of Botanical Dimensions, a tax-exempt, nonprofit research botanical garden in Hawaii devoted to the collection and propagation of plants of ethnopharmacological interest. In California, he divides his time between writing and lecturing.

RUPERT SHELDRAKE

Rupert Sheldrake was born in Newark-on-Trent, England, in 1942. He studied natural sciences at Cambridge and philosophy at Harvard, where he was a Frank Knox fellow. He obtained his Ph.D. in biochemistry from Cambridge in 1967. In the same year, he became a fellow of Clare College, Cambridge, where he was director of studies in biochemistry and cell biology until 1973. As a research fellow of the Royal Society, he carried out research at Cambridge on the development of plants and the aging of cells. From 1974 to 1978, he was principal plant physiologist at the International Crops Research Institute for the Semi-Arid Tropics (ICRISAT) in Hyderabad, India, where he worked on the physiology of tropical legume crops. He continued to work at ICRISAT as consultant physiologist until 1985.

Sheldrake is the author of *A New Science of Life*, *The Presence of the Past*, and *The Rebirth of Nature*. He is married to Jill Purce, has two young sons, and lives in London.

AUDIOTAPE AND VIDEOTAPE INFORMATION

Videotapes of the first four trialogues are available from:
Workshop Videos
620 Almar Ave.
Santa Cruz, CA 95060

Audiotapes of all the trialogues are available from:
Dolphin Tapes
P.O. Box 71
Big Sur, CA 93920